Also by Joyce Anne Trebilco:

Christmas Musings:
To Illuminate What We Celebrate

Not An Empty Promise

Joyce Anne Trebilco

WestBow Press books may be ordered through booksellers or by contacting:

WestBow Press
A Division of Thomas Nelson
1663 Liberty Drive
Bloomington, IN 47403
www.westbowpress.com
1-(866) 928-1240

ISBN: 978-1-4497-4777-0 (sc)
ISBN: 978-1-4497-4776-3 (hc)
ISBN: 978-1-4497-4778-7 (e)

Library of Congress Control Number: 2012906729

Printed in the United States of America

WestBow Press rev. date: 04/24/2012

To my family

Contents

Acknowledgments

Sometimes when I have shared one-on-one with friends about my everyday life experiences as a missionary in places where we served, Vietnam, Indonesia, and even California, people have responded by saying, "You should write a book!" Our daughter, Jeanette, has especially often urged, "Mom, you should write a book!" I am extremely grateful for these requests and votes of confidence as well as for the inspiration from such sincere interest, which certainly motivated and encouraged me to write this book. Thank you!

I am also grateful to my husband, Oliver, for his valuable editing, his keen attention to detail, and his computer savvy. His patience and enthusiastic support have blessed and encouraged me abundantly. Thank you!

Mostly, I am grateful to God for helping me recall and savor once again with incredible delight and awe these experiences of His gracious care. Thank you!

Introduction

I was a career missionary for fifteen years in Vietnam during the war there. When the Communists from the North overthrew South Vietnam in the spring of 1975, we were forced to leave and were then assigned by our mission to serve in Indonesia for six years and then in California to minister among Southeast Asian refugees. Of course, we had many interesting, even strange, certainly odd, definitely frustrating, and sometimes dangerous experiences. Yet there was for me an underlying sense of—dare I call it—high adventure that can only be accounted for by my certain conviction that Jesus Christ had called and sent me to be a missionary. But even more than that was His incredible promise: "Lo, I am with you always even to the end of the age" (Matthew 28:20b). The adventure was in witnessing firsthand, with gratitude and awe, His faithfulness in keeping that astonishing promise.

Although *Not an Empty Promise* is about my life as a missionary, it is not primarily about the missionary work we were sent to do. We have shared that many times in many meetings in churches and homes through the years, which were more or less opportunities to share with friends and family the big picture of our experiences of bringing the gospel to the unreached. Basically, they were reports of our missionary work.

But there was neither time nor opportunity to share the other aspects of our lives, which might be considered the little picture. How did we know we were called? What was it like living in foreign countries? How did we learn three languages? Why were we there? What did we accomplish?

These days, I often find myself reflecting upon and delighting in the small and not-so-small ways that Christ's promise was faithfully fulfilled. By His being with us, I do not mean that all was well, no matter what. Sometimes things were dangerous and frightening, and there were even some very sad endings. Most sad of all was that there were missionaries who were killed in cold blood in Vietnam. Was He with them? Was Christ with us in the bunkers? Was He there when bullets were whizzing around? Was He there when tribal villages very near to our house were being burned to the ground? Was He with us during serious illnesses, in our uncertainties, in our devastating grief as well as in the times of joy while living in foreign countries?

Yes, He was there with us and kept His promise in astonishing ways. It has been a blessing to me to recall and share some of them with you. All Bible quotations are from the NKJV Bible.

Ink on My Dress

A new dress! How special! I was thrilled because this did not happen to me often since my parents. Ernest and Helen Hedwall, divorced the year before, when I was nine years old. That resulted in some instability and a vague sense of being somehow different, because coming from a divorced family in those days seemed to cast a shadow on all of us. We were poor, but I didn't realize it. Probably because—as my oldest sister, Gloria Born, reminded me—"A lot of people back then were poor." I think we must have been poorer than some, because our church friends would bring us food and gift baskets at Christmas. My mother brought all seven of her children through some tough times and faithfully sent us all to Sunday school and church.

I loved my new dress. Surely there had never been a prettier red-and-white one. It tied in a perky bow at the back. I wore it to Sunday school and church. Mom wisely told me she wanted me to keep that dress to wear only on Sundays, as money was not readily available to our family in those days. However, on Monday morning, I begged and begged Mom to let me wear it to school. She reluctantly relented, and I was elated. It was going to be a really special day.

In fifth grade, when we were learning penmanship, we used scratch pens dipped in very black ink. Somehow I managed to get some blotches of ink on my new dress! That was a real disaster! I was sad that I had worn my lovely dress to school and sad that it was obviously ruined. Then there was my mother to face!

When I got home, I changed my clothes and hung the dress in the closet. Then I knelt by my bed and pleaded with the Lord to remove the ink spots. After all, I had learned in Sunday school that God hears and answers prayer, which seemed to mean that He always did what we asked Him to do. When I think of it now, I realize that I had never actually seen anyone kneel to pray. To my young mind, this would obviously be the position to take for a really serious request! So I prayed and prayed. Periodically, I jumped up and checked my dress in the closet. Alas, the spots were still there!

Eventually, I had to show the dress to Mom. Her reaction was a calm thoughtfulness, which was a relief to me. A few days later, she produced a special solution she had purchased and put some drops on the ink spots. Like magic, the spots disappeared right before our eyes. I was amazed! I was relieved! I was overjoyed! I was thankful!

I wonder now how I was so quick to think that God would, or even should, remove the ink spots from my new dress those long years ago. Why did my thoughts turn so quickly to God in my obviously hopeless dilemma? I believe it was the quiet yet strong influence my maternal grandmother, Annette Johnson, had on me when I was a small child. She made me God-conscious even though she never once spoke to me about God or spiritual things.

Grandma was born in Sweden and immigrated to America with her parents and siblings when she was five years old. As a child, I was blessed by her ladylike mannerisms and gentle ways. She had bright blue eyes that shined through her thick glasses, and her gray hair was parted in the middle with a wave on each side and pulled back into a braided bun. She looked like the grandmother one sees depicted in old storybooks. She came to our house every week and helped our mom with the wash. I also often spent time with her at her house.

Her Sunday go-to-meeting dress was black, as were her laced shoes with two-inch heels. I would watch with fascination as she secured her black, flat-brimmed hat with a very long hatpin on her head just as we were ready to go out the door to catch a streetcar to go to church. Her everyday attire was one of her many colorful print dresses, over which she always wore a large, bibbed apron. She was one neat lady!

Although she lived in the city of Minneapolis, her house, which was built by my grandfather, had never been totally modernized. There was a pump on the sink that efficiently supplied water with up-and-down movements of its handle. She did all of her cooking and baking on a wood-burning stove. There was a gas chandelier over the big, round dining room table, which never ceased to fascinate me as I watched her pull it down to almost table level, light the several globes with a match, and raise it again. I think she must have been about the last one in the city to get electricity.

Christmas at Grandma's house was a wonderful celebration for me in my childhood. Grandma always had a tall, nicely decorated, real tree that barely fit in the corner of the dining room, right next to the pump organ. The special part was that there were little white candles in candleholders clipped onto the branches. At just the right moment, when we were all seated around the dining room table, and one of my uncles had a wet towel and a bucket of water at the ready, the candles would all be carefully lit. We kids and all the grown-ups would sit in quiet wonder and watch the candles glow and flicker in the dark. There was a candle designated for each person present. The idea was to see which candle lasted the longest. This special and beautiful tradition is still observed by my youngest sister, Kathy.

I never once remember my grandmother talking to me about God or Christianity in all our times together. She was an extremely private, very quiet, and incredibly shy woman. So how is it that she made me God-conscious when I was very young? I am convinced it was because of her consistent practice of reverently bowing her head, tipping it slightly to one side, closing her eyes, and with her hands folded on her lap and her face a study of serenity, silently giving thanks to God without fail at every meal. I always peeked. I am glad I did, because I can still picture clearly in my mind her sincere, humble gratefulness. That quiet witness of her awareness of not only the reality of God but also her reverent acknowledgment of His real though invisible presence made a deep impression upon me. In fact, just being with her somehow inexplicably evoked thoughts about God.

Then there was also that oval-framed picture hanging on her bedroom wall of an angel helping two young children cross a small, rickety bridge over a deep ravine, which spoke of her faith. She never talked about such things, yet it was she who made me realize when I was very young that there was a God. My grandmother loved Him, and she experienced His presence in her life. She did not influence me with words but with quiet actions.

In my young years, a highlight of being at Grandma's was that every day she would let me feed the squirrels in her backyard. I would take the pieces of bread she gave me, throw them out, and it was fun to watch as the creatures eagerly devoured them. I remember that there was a handmade wooden number puzzle kept in one of the cubbyholes of her organ. Sometimes I asked her permission to play with it which was a pleasant diversion.

In my sophomore and junior years, my high school was only three blocks from Grandma's house, and I often dropped in to see her before taking the bus home. She would serve me a snack, probably a dish of canned peaches and a piece of her delicious homemade brown bread. We would sit and talk, but in reality I guess I did most of the talking. She always listened with obvious interest as I told her all about my high school life, which was outside the scope of her experience, as she had never attended high school. I would explain to her some of what I was required to do for homework and share with her some of my experiences. Sometimes she would exclaim in her dear, Swedish accent, "Vell, for land sakes!" That was her favorite expression and pretty much covered any situation where she was impressed, amazed, concerned, or even indignant. It was her intonation that revealed her meaning.

No, it was not Grandma who explained Bible truths to me. Yet I did learn them. At an early age, I was taught the Scriptures and encouraged to memorize verses at Sunday school and church. The Bible stories from the Old and New Testaments enthralled me. Sunday after Sunday, godly men and women taught us from God's Word, where a large group of us children sat on small, wooden chairs in rows in our junior church.

I remember staring at the picture that hung on the wall there, Alford Soord's *The Lost Sheep*, which depicted a shepherd with a crown of thorns glowing like a halo on his head. With his crook in one hand that he is using to grasp a small ledge, he is reaching down the side of a steep cliff with his other hand to grab a lost and frightened sheep, just in time to save it from a hovering bird that is ready to snatch it up. How I loved that picture! I still do. One just like it now hangs in my office. It reminds me of my childhood and of those who taught me from the Scriptures, introduced me to the Great Shepherd, and taught me about prayer.

I have often smiled at my childish yet certainly sincere prayer attempts to get God to remove those disastrous ink spots from my lovely new dress. In retrospect, though, that long ago incident did end positively. The ink spots were removed. To me, it has become a valuable lesson about prayer. God does answer prayer but not always and perhaps not usually in quite the way we hope or expect.

The Prodigal

The meeting had started. Good! At the large First Covenant Evangelical Church in downtown Minneapolis, ushers had even brought in extra chairs and placed them in the aisles. It was going to be a good meeting. Like the tide that washes over the sand, a warm wave of contentment surged over my heart. I loved being in church. It was April 23, 1948, and I was fourteen years old.

The radiant, cute song leader announced the page number, and his enthusiasm was contagious. Soon the singing was over. Then the serious, young evangelist, who was the guest speaker for that week of special meetings, stood up and stepped to the pulpit. I was sitting in the front row, just above the clock, in the middle of the U-shaped balcony with a perfect view of the pulpit. My twelve-year-old brother, Gary, sat next to me. I gazed around, looking for familiar faces and trying to spot the visitors.

The evangelist's voice brought my attention back to the platform. "Please turn in your Bibles to Luke 15." I found it. Oh yes, the story of the prodigal son. He read the passage. I began thinking about my schoolwork and how busy they kept us in ninth grade. Happily, I had finished my homework before coming to the service.

"Listen!" the speaker commanded. I began to pay attention again. He was relating how the father in the story obliged his youngest son by giving him his inheritance. Then he told about that son leaving home, caring nothing about his father's disappointment and broken heart. Out he went from his home with his share of the inheritance that he had requested, off to the big city to have his fun. He was so happy, so free. He was going to have a great time. Wonderful bliss! He was on his own at last. He found new like-minded friends and enjoyed a life of revelry, which lasted as long as his money did. When his money was gone, his life took an awful turn. His friends deserted him.

Suddenly, it was as though I were no longer in church. I was there with the wandering boy. Now he was destitute. He was unhappy, penniless, and friendless. Finally, he got a miserable, lowly job feeding pigs. He felt so alone and was so hungry. Looking at the pig food, he could have eaten it, as there was a famine in the land. He had been hungry for a long time since he had wasted his money.

How many times had I heard that story? Scores of times, I suppose. I had never once thought of myself as a prodigal, though. I had usually aimed to please, to obey, and to be good. Frankly, I was not like this desperate boy. No, I would definitely not have been considered a prodigal. The evangelist continued, "Oh, my friends, the famine of the boy was not just physical; the real famine was in his soul!" Famine! The word blazed in my mind and penetrated my heart. Suddenly, I felt a gnawing famine at the very center of my being, and I became extremely thoughtful. God the Holy Spirit probed deeply. Somehow it became clear to me that my goodness was unclean and filthy before a holy God. Even though I was religious and loved being involved in church meetings and activities, I began to realize that I had an unmet hunger within myself. It was spiritual famine.

The evangelist then described how this wayward son suddenly remembered the good life he once had back home. The story continued. The boy came to himself. In other words, he suddenly realized that the pleasure he once craved did not bring satisfaction. He must have been out of his mind to leave his loving home. Suddenly, he made a decision right there in the pigpen: he decided to return to his father, ask for forgiveness, and humbly plead with him to be allowed to become a lowly servant. He obviously did not understand his father.

Well, the wonderful story unfolded, and there was a tremendously happy reunion between the boy and his father. Actually, his father must have hoped that one day this would happen because he seemed to always be watching for him, and when he saw his son coming down the road toward him, he ran to meet him. The son's request to be a servant was rejected. The father ordered the servants to put sandals on his feet, give him new clothes, and then the ring with the family seal was slipped on his finger. A feast was prepared to celebrate the return of this son. Oh, the joy, the peace, and the amazement the son must have experienced as his father lavished him with undeserved, unconditional love and forgiveness.

I sat stunned by the description, knowing that I was still unclean and starving back there in the pigpen. The preacher extended a clear invitation to come to God the Father to be cleansed. This could only be possible because Christ had shed His blood on the cross and had been my substitute for the punishment I deserved because I was a guilty sinner. But I was way too shy to go forward in that large, packed church. I knew I could never walk from the balcony all the way down the aisle of the main floor past all those people and kneel at the altar. My legs were as lead. My heart was thumping. It was too hard! It was too far! I couldn't do it, I couldn't!

But I did! On the last line of the last chorus of the last verse of the invitation song, I put my hymnal down. The first step was the hardest. I made my way down to the main floor, falteringly at first, but then boldly. I walked past all those people right to the altar where I slipped to my knees. Eleanor Hedlund, a dear friend and servant of the church, knelt beside me and put her arm around me. A few minutes later, I stood up. My heart was full. The rags were gone. The famine was over. I had been welcomed into the family of God.

Back home, I opened my Bible to my daily readings. Somehow, someone had gotten through to me as a teenager the importance of daily Bible reading, for which I am immensely grateful. That night, I read these words: "As Jesus passed on from there, He saw a man named Matthew sitting at the tax office. And He said to him, 'Follow Me.' So he arose and followed Him" (Matthew 9:9). Two words seemed to jump off the page: *"Follow Me."*

Startled, I finally realized it was His message not only for Matthew but also for me! Jesus was talking to me! So began my precious experience of being personally fed and challenged by

Scripture, which is when my daily Bible readings for the most part changed from duty to desire; from having to, to wanting to.

God's inspired Word is meant to instruct, reprove, and correct, and also to refresh, comfort, and strengthen me. And so it did. And so it does.

Summer Missionary Conference

When I graduated from high school in the spring of 1951, I was so concerned about my future. That night, I knelt by my bed and pleaded with the Lord to somehow send me a list of things that showed me His plan for my life. He never got that list to me, because following Him is all about faith and learning to take one trustful step at a time.

At the end of that summer, I learned there were secretarial jobs available in public schools, which required taking a civil service test. I was so happy when I got a job as the clerk/stenographer at Hiawatha Elementary School in Minneapolis. I loved that job because there was so much variety, serving six hundred children and twenty teachers. I had a great principal, wonderful teacher, and parent friends. With part of my salary, I was even able to help with mortgage payments on our family's newly purchased house, which I was glad to do so since I had helped talk my mother into buying rather than renting.

In Sunday school, I had become acquainted with Delores Johnson, (who later married David Shaver), with whom I became good friends. We had a lot of fun together and had our own jokes. Since both of us were a little chubby, we had this saying, "Some girls have it, some girls don't have it, and some girls have too much of it!" Oh well, we thought it was funny. We laughed a lot. Once we went on a fifteen-mile bike ride to my sister Gloria's house, which was quite an accomplishment with its up-hill challenges on our one-speed bikes. But we made it!

Our friendship was not just about fun, though. We did a lot of heart sharing, and I was often impressed with how much Delores wanted to please the Lord. She was the first person I ever knew who had tears in her eyes when she felt she had failed Him. How true it is that some people come into our lives for just a short period of time and then leave. Others stay and impact us in such a way that we are never ever the same. Delores was that kind of friend to me.

In the summer of 1953, Elsie Wold (later married to Fred Johnson) invited Delores to attend the annual Missionary Conference at Bethany Fellowship Missionary Training Center in Bloomington, Minnesota. Delores was not interested and did not want to go. However, Elsie kept bugging her to attend, and Delores kept bugging me to go with her. Reluctantly, we gave in and went. The city bus did not take us to the entrance of Bethany, so we had to walk a short distance after getting off the bus. It started to rain, and we got soaked. By now, I was pretty grumpy. But we were finally able to dry off and get fairly presentable for the meeting that evening.

When we were seated, I was completely unaware that I was about to hear a life-changing message. The Reverend Leonard Ravenhill from England was the speaker. His message was: "Or do you not know . . . you are not your own? For you were bought at a price" (1 Corinthians 6: 19b-20a).

It is hard to describe the powerful impact this had on me. You might say the message was a scathing, straight from the hip, this-means-you kind of presentation. I needed to face the truth of what commitment to Jesus Christ really means. Five years prior to hearing this message, at the time of my conversion, I had sensed that God wanted me to be a missionary. In fact, China came to mind. However, I had long ago dismissed the idea. Now, though, I could not ignore that God was asking something of me. It was (oh no!) to quit my job and go to Bethany.

There was a struggle. Some of my coworkers and even some of my friends were definitely not impressed with my thinking of going to Bethany. Nevertheless, in the fall of 1954, I enrolled in their three-year program. I was interested in getting Bible training but not particularly thinking of becoming a missionary.

I hesitated leaving my job because of my promise to help pay our family's mortgage payments. But God's timing was perfect; just about then, my family's financial situation had changed enough so that my mother gladly freed me from any financial obligation and sent me off with her blessing.

Quitting my job was a big step for me, but I knew without doubt that God was leading me. I was content yet naively ignorant of where that one step of obedience would eventually take me.

My Dorm Room Chair

My first roommate was Meredith Nelson. I remember being dismayed to learn that she had just graduated from high school. After all I was a young adult who had worked three years after graduating. I learned quickly, though, that maturity is not always measured by years, especially spiritual maturity. That was the first of many lessons I would have to learn. Meredith and I hit it off and had good times together, especially as we had long discussions about life and the Bible. I am thankful for that one year we had together, as roommates and for our ongoing friendship.

It was a special honor when I was invited to be to be the bridesmaid in my friend Priscilla Newton's wedding. She married Dick Dugan. We were classmates. Dick's brother, Leroy Dugan was one of my favorite teachers at Bethany and also my boss in the composing room of Bethany's small print shop. It was some years later when that print shop had grown to become the large and successful Bethany House Publishers.

One night during my junior year at Bethany, I could not sleep. By then I knew that God wanted me to be a missionary. My thoughts were in unending confusion and turmoil about all sorts of concerns that were of paramount importance to me. To which foreign field was the Lord going to send me? How could I ever learn a foreign language since I had already decided I couldn't? Where would I be sent? How was I ever going to get the finances I needed? Could I really be a missionary? What about marriage? On and on it went. I discussed these concerns over and over again with myself, always with the same weary, dreary apprehensions. Finally, I recognized that the whole thing was too big for me, and that I did not trust the Lord for any of it. Something was definitely wrong!

I dragged myself out of bed with a heavy sigh. Picking up a small notebook, I crossed the room and knelt by a chair. Opening the book, I found a page on which I knew I had written a simple yet challenging outline of a sermon I had heard. Gradually, I began to realize some things.

While I was working as the clerk/stenographer at Hiawatha Elementary School in Minneapolis, I felt like I was in charge of the whole school, and in many ways I was, and I loved it! I took dictation in shorthand from the principal, typed his letters, and was in charge of the mail. I was in charge of filling book and supply requisitions for the teachers, delivering everything to their classrooms, duplicating pages and pages of student worksheets, preparing the weekly teachers' bulletin, operating the switchboard, answering the phone, compiling the required attendance and financial reports for the main office, sometimes taking children's temperatures, applying Band-Aids, decorating and updating the bulletin board, and making the coffee and serving sweet rolls to the staff every morning. I thoroughly enjoyed this variety of responsibilities. And of course, for all this I earned a salary. Now, however, everything was different, as much had changed. A gigantic concern was my struggle to really trust the Lord.

As I knelt there, I became aware of a previously unrealized core hesitation in my heart to surrender completely to God's will for my life. The argument could be made that since God is the God of the universe, He is already completely in charge. Surely, though, God desires voluntary obedience, which I knew meant giving Him all I was, all I had, and all I hoped to be. That certainly covered everything and meant my life would no longer be about what I liked or did not like, what I wanted to do or did not want to do, where I wanted to be or did not want to be.

I was thoroughly convinced of God's love for me, but somehow I was not sure I could trust Him to actually be completely in charge of my life. In a way, I felt I had already taken a huge step of obedience. After all, I had given up the job I loved and was in Bible school. Did

I somehow think that this equaled sacrifice? Perhaps I did to some degree, and if so, how shamefully foolish. Years later, I learned that David Livingstone, who suffered so many severe trials in opening Africa to the gospel, said: "I do not mention these privations as if I considered them to be sacrifices, for I think that the word ought never to be applied to anything we can do for Him, Who came down from heaven and died for us." Of course, I never could come anywhere near experiencing any of the hardships that David Livingstone suffered, yet I did eventually learn that the word *sacrifice* must never be applied to anything I could do for God.

Kneeling by my chair I began to realize I had not really submitted to the lordship of Jesus Christ as a wholehearted disciple but instead a reluctant one. All of my worries revealed my lack of surrender and especially my lack of trust. Must there always be this struggle, this fear, this hesitation connected with obeying and trusting Him? Was that the way it was supposed to be for Christians? Jesus said, "But why do you call Me, 'Lord, Lord,' and not do the things I say?" (Luke 6:46). This surely implies that He expects His lordship to be honored by us. Jesus also said, "If you love Me, keep My commandments" (John 14:15). If He expects willing even joyful obedience, it surely must be possible.

> I beseech you therefore, brethren, by the mercies of God, that you present your bodies a living sacrifice, holy, acceptable to God, which is your reasonable service. And do not be conformed to this world, but be transformed by the renewing of your mind, that you may prove what is that good and acceptable and perfect will of God.
>
> —Romans 12: 1-2

Willing submission would certainly be exemplified by peace and joy and not by a begrudging attitude and constant struggle, self-pity, and even fear. For example, it would be odd if in a marriage relationship the wife kept her marriage certificate handy because of her need to remind herself that she was married and had to be a wife. What kind of wife would that be? There would certainly be no spontaneous joy, only a cold sense of duty, especially during difficult times. I have heard some Christians tell of their stubborn resistance when they realized what God wanted of them. One man even said, "God threw me to the ground and put a sword to my throat!" To me, that sounded like what he was really saying was that when he finally obeyed, it was because he felt forced, almost like, "God made me do it." What kind of disciple would that be?

My desire was to have a consistent Christianity that would experience joyful obedience and peaceful trust. So why was I so hesitant and doubtful? There were plenty of excuses. One could reason that my hesitation was probably not really my fault because I couldn't help being distrustful since I was living in a fallen and broken world. Or I could always blame the Devil, something from my broken-home childhood, or just plain human frailty. However, being intensely exposed to Bible truths day by day, I had already come to realize that nowhere does Scripture condone the blame game. "Each one is tempted when he is drawn away by his own desires and enticed" (James 1:14). I realized that even as a Christian, self-seeking and self-love were my problems. I was used to running my own life and loving it! Everything was more or less all about me.

At Bethany, I was privileged to hear daily the need to relinquish it all, to die to self and thereby be identified with the cross of Jesus Christ. Every day at Bethany I was challenged by message and also by godly example to realize that the Christian life was to be all about Him.

Still kneeling by my chair, I began to pray, and my prayer was simple. I willingly and firmly gave up my right to my life and let Him have it all. And I really meant it. I yielded to His plans for me, whatever they might be, and by faith did some serious reckoning!

> Likewise you also, reckon yourselves to be dead indeed unto sin, but alive to God in Christ Jesus our Lord. For sin shall not have dominion over you . . .
>
> —Romans 6:11, 14

> I have been crucified with Christ; it is no longer I who live, but Christ lives in me; and the life which I now live in the flesh I live by faith in the Son of God, who loved me and gave Himself for me.
>
> —Galatians 2:20

> For the love of Christ compels us, because we judge thus: that if One died for all, then all died; and He died for all, that those who live should live no longer for themselves, but for Him who died for them and rose again.
>
> —2 Corinthians 5:14-15

I got up from my knees by the chair that had become my altar of surrender. My choice was clear, my heart at peace, and most incredible of all, my mind was quiet. The matter had been settled. I had surrendered to His lordship over my life. I will never forget this because I was subsequently aware of a welcome sense of relief—tremendous relief! I had come to the end of my puny, hoarded resources and was at the beginning of God's yet-untapped abundance. I became aware that my life would now be about who He is and what He could do and not about who I am and what I cannot do. It was a no-turning-back choice, a moment of heartfelt decision, a done deal that would steady me many times in the years ahead. Now, with every situation that required obedience, there need be no unnecessary debate. I had already purposed in my heart and was fully convinced that He would provide the needed grace. I will always be grateful for that cherished night. I went back to bed and enjoyed a delightful, peaceful, restful sleep.

I have often noted that a small child will look into the face of a parent when there is something happening that seems scary or questionable. Why is that? It obviously helps the child to evaluate the situation and get needed reassurance even though he or she still not does fully understanding the situation. How wonderful that from now on I could be like that, and let Him do the worrying and the planning for my life. Most surprising of all, He didn't need my help! He would be in charge. He just required my obedience. Since I had surrendered to Him and pledged my willing obedience to His will, it was His responsibility to enable, guide, lead, and provide.

I became profoundly aware that He should and could be trusted. This was not a new truth to me, but it had now become internalized, personal and authentic. Without question, something had changed, because I now began to find it easier to look with childlike trust to my heavenly Father for needed reassurance. Someone once said, "God does not hurry us, but He does wait for us."

We read in the Bible of Daniel, a handsome, young, intelligent Hebrew lad of noble birth who was taken with others from Israel as a slave into the palace of the king of Babylon where they would be taught the Babylonian language and literature. Daniel, a devout Jew, had purposed in his heart, though, that he would not succumb to the heathen culture of the Babylonians. "But Daniel purposed in his heart that he would not defile himself" (Daniel 1:8a). Surely this beforehand resolve was the strong defense that kept Daniel true to his God throughout his whole fantastic, godly life.

There would always be my need of God's ever-available grace, strength, and guidance. Yes, there would be challenges, important choices, and many difficult decisions ahead. Commitment is surely not worth much until it is tested. My focus was not about being perfect but about relinquishing my will to Christ's will as my Lord and Master. I knew there might be times when I would stumble or even fall (and I did), but that His gracious forgiveness would always be available. Also available was Christ's victory over sin and Satan, which was accomplished through His death on the cross and His resurrection.

Scripture indicates clearly that our expectation should be to walk in triumph and not in the expectation of defeat. I am so grateful that God led me to Bethany because there I became keenly aware of and was profoundly blessed by exposure to the message of the cross. The victory Christ won for us on the cross was accomplished to give us a life-changing victory, a triumphant victory through His name.

Now thanks be to God who always leads us in triumph in Christ, and through us diffuses the fragrance of His knowledge in every place.

—2 Corinthians 2:14

Now unto Him who is able to keep you from stumbling . . .

—Jude 1:24

Yet in all these things we are more than conquerors through Him who loved us.

—Romans 8:37

I can do all things through Christ who strengthens me.

—Philippians 4:13

Now to Him Who is able to do exceedingly abundantly above all that we ask or think according to the power that works in us.

—Ephesians 3:20

Walk in the Spirit, and you shall not fulfill the lust of the flesh.

—Galatians 5:16

Walk as children of light.

—Ephesians 5:8b

Walk worthy of the Lord, fully pleasing Him, being fruitful in every good work . . .

—Colossians 1:10

For we walk by faith not by sight.

—2 Corinthians 5:7

> And my God is able to supply all your need according to His riches in glory by Christ Jesus.
>
> —Philippians 4:19

Though the journey ahead was still unknown, and perplexing to me, I finally realized that it had to be traveled by faith, taking just one step at a time. That, of course, is the essence of faith. He could enable me to grow in grace and become what I should be and what I longed to be—His willing, obedient servant. It was a peaceful new beginning for me to learn to follow Christ unencumbered by competition between His will and mine. There would be much for me to learn, but now I felt teachable and eager.

> Not that I have already attained, or am already perfected; but I press on, that I may lay hold of that for which Christ Jesus has also laid hold of me.
>
> —Philippians 3:12

The downward tendency always tugs on us in our broken, fallen world. Yet I experienced a heartfelt sense of release and rest from my pushing against the leading of God's Spirit and reluctance to even know His will and went on to experience a growing sense of cooperating with Him on a daily basis. I was grieved to realize that I had been on the defensive against God, but now that was no longer the case. I had crossed over into a relationship of trust, realizing as never before that He is the One who is totally trustworthy. I could and would trust Him, no matter what.

A lady once told me that she could never fully surrender to God for fear of what He might ask of her. She was right in one sense, because He does ask a lot, in fact He asks everything! That's why we call Him the Lord Jesus Christ. He is *meant* to be our Lord and Master. But her fear, though understandable, was certainly a sad commentary on her perception of our loving heavenly Father. She obviously felt that God would somehow hold out on her.

It was for that exact same reason that Eve succumbed to Satan's cunning craftiness in the garden of Eden, where he planted seeds of doubt as to God's goodness. She listened to him, believed the lie, and thus was finally and fully deceived. She became convinced that God was keeping something good from her, so she would find out for herself and willfully and willingly ate the forbidden fruit. By choosing against God, Eve had submitted to Satan's tyranny. In George MacDonald's sermon entitled "Kingship," he makes a sobering comment: "The one principle of hell is, 'I am my own.'"

Accepted

After graduating from Bethany Missionary Training Center, Delores and I applied to the Worldwide Evangelization for Christ, an international, interdenominational missionary organization. We were accepted, and in January 1958, we traveled by train from Minneapolis to Fort Washington, Pennsylvania. Although I knew that WEC headquarters was once the mansion of a wealthy family, I was pleasantly surprised when I actually saw it. The headquarters was on a hill that we reached by a curved road sided by low, thick stone walls through a wooded area. Suddenly there it was, right in front of us—the beautiful castle-like mansion. It seemed like we had arrived in England. The castle had forty-three rooms and thirteen fireplaces. We were taken to our dormitory on the third floor. Soon a bell rang, announcing the evening meal. Our new life had begun.

Back in those days at the headquarters, there were several British people on the staff who influenced the place and caused us candidates some confusion. Tea, we learned, was what some called the evening meal. Hot tea was definitely the beverage of choice at WEC and was the one most available. That was quite a change for me because in my Scandinavian Minnesota almost everyone drank coffee. However, knowing that this was going to be the situation, I prepared myself by gradually developing a taste for tea. My mother was incredulous when I started drinking hot tea, of all things!

The British accent of some on the staff took some getting used to. The talented candidate director and his wife were from Scotland,

and their English was different, really different! Early on, when he asked us candidates to turn in our Bibles to what sounded to me like, "Sam," for a moment I wondered if he meant First or Second Samuel. But it was neither. It was Psalm. His wife, who was responsible for preparing for guests who would be attending the monthly conference at the headquarters, instructed me to put a cot in one of the bedrooms. I was amused when I discovered that she meant a baby crib and not a single-person cot, and I scrambled to fix the misunderstanding. Such instances were good for cross-cultural experiences. I realized that this was nothing compared to what I would eventually experience on some still-unknown-to-me mission field.

We missionary candidates attended a phonetics class where we began to learn to hear, mimic, and write strange-to-us consonant and vowel sounds. It was an important and valuable class to better prepare us for future language study. I enjoyed the class, and was pleasantly surprised to discover that it was not hard for me at all. In high school, I had not taken a language elective because I was convinced I could never learn a foreign language. I once heard it said that God doesn't call the able, but He enables the called. Whatever the case, I was no longer afraid of having to learn a foreign language.

We had other important daily learning experiences. Every weekday morning, the staff and the candidates met to pray together after breakfast. We read Scripture, received a word of exhortation, and letters from various mission fields from around the world were read and their needs brought into focus. Then there were the needs right there at the headquarters that required prayer. No need was too big or too small to bring before the Lord. We each had a rug mat by our chairs that we knelt upon on the hardwood floor and prayed fervently. It was an enriching experience being there with seasoned missionaries now serving on the staff and with my fellow candidate friends, who were preparing to go to various countries: the Congo, Portuguese Guinea, Columbia, Jamaica, Dominica, Thailand, and Vietnam.

We all had jobs at the headquarters. We girls were moved around to fulfill daily responsibilities, such as cleaning, cooking, baking, setting tables, washing dishes, or preparing rooms for guests. I was often assigned to baking, sometimes pies, ten or twenty at a time. It was challenging, but I loved it! One time, though, when I was making a huge batch of baking powder biscuits (scones), I mistakenly used soda instead of baking powder. Whoops! That did not go over too well, as they were horribly bitter and thus inedible.

In the midst of all this activity, we candidates were trying to patiently wait for when we would hopefully be accepted and then could move off to whatever country God had laid on our hearts, with, of course, the approval of the staff.

When I arrived at WEC headquarters as a candidate, WEC was entering a new-to-them field, which was Vietnam—more readily known as part of French Indochina. As time went by, I became more and more interested in Vietnam. Up to that point, I was unsure as to where I should go, although in my senior year of Bible school there was a clear moment when I knew God was impressing Asia on my heart. In fact, I can point out the exact spot on the sidewalk where I was between the administration building and my dorm at Bethany when suddenly the words, "I am sending you to the Orient" were clearly transmitted to me somehow. I knew then that God would send me to some place in Asia.

What a nervous thrill it was when I stood before the staff at WEC, was given the right hand of fellowship, and was considered an accepted missionary candidate for Vietnam. In September, I was introduced to all the guests who came to the monthly conference as the first WEC candidate from the US to be accepted for Vietnam. I was twenty-five.

WEC is a faith mission, which is why it appealed to me. Every person on the staff, every missionary, and every missionary candidate is committed to trusting God to supply all needs, personal and communal. All gifts are given to the person to whom it was sent or to any specific ministry, not applied to other needs or overhead expenses.

While I was still a student at Bethany, a missionary who was in the area was given the opportunity to speak to us students, as was often the case. I am sad to confess that I do not remember that person's name, but I do remember vividly his message. He described three types of the life of faith for missionaries and explained that all were legitimate: One, tell people what you need; two, pray for your needs at a group prayer meeting; and three, just tell God. Now that last one really appealed to me. Admittedly, it was not the only method, but I guess I liked the challenge.

It was not as though I could never let my needs be known, but I wouldn't be running around asking people for money. What an excellent way to live, I thought, to trust God and leave it up to Him to open the hearts and wallets of people without having to "beg." Such a lifestyle would certainly reveal His personal and practical care through unsolicited provision and would also confirm His leading and involvement. Of course, the supply would usually come through God's people but would still be clearly from Him. Receiving one dollar or one thousand dollars would inspire the same thankfulness and joy.

After a meeting I had spoken at before I went to Vietnam, an elderly lady was also getting a ride with my hostess. When we stopped at the lady's store, she asked me to come in with her for a minute. Once inside, she rang open her cash register and handed me a five-dollar bill. I will never forget that precious moment or that widow's small yet generous and much-appreciated gift.

I came to realize the importance of not only being thankful but also expressing that gratefulness to the giver and of course to God. It was always especially exciting to convey to anyone who supplied a gift that it was exactly what was needed so that they too could be awed and blessed.

Many, many years later, our ministry needed a new laptop computer. On a Monday, we went to price one. Even on sale, it was too expensive for us. Then on Wednesday, we received an unexpected check in the amount almost to the penny of the cost of the computer we had been considering. The friends who sent the gift knew nothing of our need. No one did. And that particular group of people has blessed us many times through the years. That time, though, they were encouraged to learn of how timely and *apropos* their gift was to us.

Some call it coincidence. Someone once said, "Coincidence is when God chooses to remain anonymous!" He does not tap us on the shoulder and say, "Watch this!' No, He works without drum roll. How it must rejoice His heart, though, when we the recipients recognize His fingerprints on the quiet supply and respond with grateful thanksgiving to Him and His faithful givers.

Those many years ago in Bible school, I had wholeheartedly embraced the life of faith. As a student, I had been eager to get started. When I was accepted by WEC to go to Vietnam, though, I only had about five dollars to my name! Keep in mind that I was in Pennsylvania, had to get home to Minnesota, and then on to Vietnam, literally halfway around the world, and live there. My trust and commitment were about to be put to the test. This was it! Suddenly, there was a clashing of my intellectual faith with my circumstances. Now what? I found myself facing a very real and daunting challenge.

The Challenge

I felt an urgent need to get away by myself for some time of quietness. Dorm life does not provide such a luxury, so I asked a friend on the WEC staff if I could use her bedroom for a short time one afternoon when I knew she would not be there. She graciously consented. I took my Bible, closed the door, and knelt by her bed. I was quiet, not knowing exactly how to pray about the finances I now urgently needed.

I must have been reading the Psalms because I remember vividly that when I opened my eyes I found myself staring at Psalm 142:7b, ". . . the righteous shall surround me, for You shall deal bountifully with me." I was thunderstruck! In an instant, I jumped to my feet and said with glee in that borrowed bedroom, "Lord, You called me to serve as a missionary in Vietnam, so the ticket's on You! Thank You. Thank You!" Since going to Vietnam was His idea not mine, of course He would be the One to make it happen. I just stood there for some minutes, delighting in this fantastic promise of His supply, even bountiful supply, which would certainly be needed to get me there.

At that time, I was glad to be asked to be part of a missionary team going to New York State to minister in several churches. As a candidate, I was given ten minutes to share my testimony of how I knew God called me and was leading me to be a missionary in Vietnam. As I remember, I think I usually could give it in twelve minutes. The other speakers were seasoned missionaries, sharing about their ministries in different parts of the world. Soon after returning to WEC headquarters from that tour, I was released from all responsibilities at the headquarters and started packing my suitcases and footlocker.

October 31, 1958, was my last day at WEC headquarters. They had a tradition that I liked very much: The dinner bell was rung each time someone was moving toward his or her overseas assignment, which alerted everyone to stop work and come from all the buildings to gather at the front of the castle for a prayerful send off. I had rejoiced several times to be one of those who gathered to send someone off. Now it was my turn. I felt a warm and wonderful sense of oneness, of love and good wishes as I went around the circle expressing my thanks and saying my farewells.

A missionary couple I had not met was in that gathered group. I think they were starting their furlough. As I went around the circle, I was surprised when the husband quietly handed me a small, white envelope. Afterward, I slipped away and opened it. Inside was two hundred dollars in cash! In those days that was like a fortune to me, and it surely was bountiful! I was thrilled! As I held those wonderful, green bills in my hand, my heart was filled with wonder and joy. This surely was bountiful supply! God had so soon supplied me with this wonderful provision through this missionary couple, Marlin and Barbara Summers that would easily help me begin the long journey ahead. It was time to get started.

Having been invited again to be part of a team of missionaries who were speaking in churches in Ohio and Michigan, I got in the car, and we were off. A by-product of this road trip was that it would get me closer to Minneapolis. I had the opportunity to share my testimony many times in several churches. We moved around two by two to different churches that were having special mission-awareness meetings. We also moved around to homes that were opened to us. The hospitality of these people was gracious and generous. Not only was that wonderful but these friends were deeply burdened for the multitudes around the world who had not yet heard the good news of the gospel of Jesus Christ, which of course enhanced our fellowship together. One of those small churches decided to help with my financial support and began sending gifts each month for many years. Even now, and since 1959, we still sometimes receive gifts from them. Such faithfulness amazes us.

When these meetings were finished, I traveled by bus to Chicago, where I stayed with one of my mother's friends, a captain in the Salvation Army. From there, I planned to go home, also by bus. I purchased my ticket and checked my footlocker in at the bus depot.

I had been invited to a Christian Business Women's Christmas meeting and dinner. After the meeting, I was enjoying their dinner when a lady came over to my table and introduced herself to me. She had just learned that I was traveling to Minneapolis and said she would be delighted if I would travel with her, as she too was leaving the next day and would be driving to her family home which would take her past Minneapolis. So it was arranged, and so it came to pass. The next morning, I had time to get a full refund on my bus ticket and was told that they would ship my footlocker on the bus.

So I had the privilege of traveling with my new friend who drove me right up to my family home in Minneapolis. That was on December 20, 1958. Not only had God brought me back to Minneapolis in a wonderful way, but He also provided me with many wonderful lifelong friends and meaningful, enriching experiences along the way.

There was still the need to get to Vietnam, though. I felt confident that the One who called me to go there was the One who would make it happen, so I would just trust Him. I do not remember having any anxiety over my situation, for which I hardly recognized myself anymore. I just knew it was not my concern but His. In fact, I remember being interested in seeing how He was going to make it happen and basically sat back and watched, confident that He could and would do so.

By Freighter to Vietnam

I was thrilled to be home in time for Christmas, 1958. It was great being with my family again. I got a part-time job at the Department of Education in Minneapolis and had a lot of friends to catch up with after being away for almost a year. In the midst of my excitement, I was keenly aware that the New Year, 1959, was going to be very different for me.

I was booked to sail on the Norwegian freighter the M/S *Fernsea* that belonged to the Maersk line, which had accommodations for ten passengers. There was packing to do and a need to know *what* to pack! I had gotten some advice and worked on my list. Generously, my mother gave me a large trunk, which was perfect for packing the supplies I would be taking. She also gave me a cedar chest. I was pretty much going to set up house over there, and since I was going by freighter, everything was included in the price of the ticket, made possible from the gifts of many interested individuals and churches. My own church had started to help with some of my support, which was a tremendous blessing.

The trunk and my footlocker, were carefully packed. After the cedar chest was packed, it had to be crated, and then each piece had to be stenciled with my South Vietnam address in the city of Danang. My brothers Gary and Ken loaded it all into a trailer and took it to the *Fernsea* trucking agency, where it would eventually get to the docks at San Francisco and from there onto the freighter.

Meanwhile, I was preparing to go to San Francisco myself. There were many immunization shots I had to endure. Then there was a prayer card that had to be designed and printed. My married sister Gloria and my mom helped me with some sewing projects. It was a time of many wonderful opportunities to speak at meetings and to enjoy the several showers that friends were having for me. I was also busy writing and sending thank you letters to individuals and churches for their gifts, setting up a WEC display at a local Bible school, writing a prayer letter, getting my visa and passport, and also trying to spend as much time as possible with family and friends.

It was at this time I had to learn an important lesson: patience. I was so ready and eager to go. I had said good-bye to many people and was booked to sail on April 11, 1959. However, a problem had developed. In those days, a single girl was not allowed to travel on this freighter unless there were other lady passengers. There weren't, so I had to re-book for May 12. It was not much fun arriving at church for another month and have people exclaim, "Are you still here?" This required an explanation over and over again. Do I know why the wait? Not really, except it gave me extra time to be with my family. When I finally did sail, though, I met some wonderful, interesting people.

God was so good to me. My mother, stepfather, and my fifteen-year-old sister, Kathy, were going by train with me to California, which was pretty special. Also, the trains back then were real trains and wonderful with their big windows, tables with comfortable, bench-like chairs, and an elegant dining car. It was a fun trip across this wonderful country. We spent some time with our California WEC representative and his wife. Finally, it was time to board the ship that would take me to Vietnam. I was given a double stateroom to myself because they were one passenger short. Imagine that! A spacious room by ship standards, the sitting room area had a couch, an armchair, and a coffee table positioned under two curtained portholes. My folks, sister, and I sat together there for awhile, but it was soon time for them to leave. When they left, I cried.

I set sail at five p.m. on May 14. I was terribly seasick for a couple of days but eventually got my sea legs. I discovered that traveling by freighter was really enjoyable and the perfect way for me to relax from all the busyness I had been experiencing. It also gave me time to remember all the ways God had led and blessed me up to this point, which encouraged my faith in Him for whatever I would face in the future. I was excited. Vietnam, here I come! Back then, before the war there, people did not know much about Vietnam. But they would soon learn.

We enjoyed sumptuous food, elegantly served to us at the captain's table. Afterward, we were invited into the parlor to enjoy Norwegian coffee, chat, and sometimes play table games. While reclining on a deck chair, there was plenty of time to read, enjoy the wonders of the endless sea, be amazed at the sunsets, and be awed by the star-studded sky. I played shuffleboard occasionally with other passengers. The captain helped me send my mother a birthday telegram from the ship, which I thought was pretty neat. Ever so gradually, I was moving toward the country that would be my home for many years.

We arrived in Manila Bay, Philippines, on June 3, where I was kindly met by missionaries whom WEC had informed of my arrival. The ship would be in dock for a couple of days, so they graciously invited me to stay with them.

For the first time, I saw a one or two friendly geckos, those small mosquito-eating lizards, walking on the walls and ceilings of my bedroom. Also, with the exception of the missionaries, I couldn't understand a word people were saying. And in Manila, I had my first taste of the delicious mango fruit at the home of a missionary couple from my church who were serving in Manila and had invited me over for a meal.

All too quickly, I had to return to the ship. Our next stop was Hong Kong. An elderly, well-traveled widow on the ship and I had become friends, and she wanted to show me the sights of Hong Kong. What a city! She took me to many interesting places, and I shopped for gifts for upcoming birthdays in my family. This was truly Asia, a fascinating and bewildering experience—like China Town in San Francisco but ten times more so, as it was the real thing. In the evening, my new friend invited me to her room of the Peninsula Hotel, where she ordered dinner from room service, which we enjoyed together. The evening view of all the lights on the buildings and the ships in the harbor from her hotel window was fantastic. It had been a most enjoyable day for me, but then it was time to say farewell to my friend and return to the ship.

We sailed the next morning. Now I was the only passenger on board. The crew did not speak much English. Even the stewardess didn't seem to know much. The captain did well understanding English and spoke it with lovely Scandinavian accent that reminded me of my dear Swedish-born grandmother.

We arrived in Saigon on June 12. I leaned over the railing, fascinated as I gazed down at what looked like a sea of cone-shaped, straw hats, under which people were busily moving about on the pier. The captain joined me and told me that he could not understand why I, a single girl, would ever want to come to Vietnam and live in this foreign country. He was shocked when he realized I was so happy to have arrived. I could understand how he felt. He, of course, did not realize who it was who had sent me, and that I had come in obedience to God's clear call upon my life.

I noticed a gentleman coming up the gangplank. Milt Barker, a missionary from another mission, introduced himself, and I was relieved to meet him. He had been commissioned by WEC Vietnam to meet me, help get me through customs, and then settle me in a hotel, The Saigon Palace. This he kindly did, and I was most grateful.

That night, I crawled into bed under the mosquito net (a new experience) and was soon fast asleep, though still somewhat dazed, amazed, and grateful that I had actually gotten this far. The next day, I wandered around an open market and heard the Vietnamese language for the first time. Who could ever learn such a language? This language with its six tones was definitely going to be a challenge, but I had the firm conviction that God's enabling, the prayers of many friends, and my hard work would be the necessary ingredients for my success.

Soon it was time to go to the airport. The hotel clerk called a taxi for me and told the driver where I wanted to go. After checking my baggage, I stood with a crowd of people on the runway, waiting to board a small plane that would take us to Danang. A trim stewardess with clipboard in hand stood at the bottom of the steps leading up and into the plane. She motioned for me to come forward and had me board the plane first. I was surprised and somewhat embarrassed but obediently followed her up the steps into the plane. It seemed like colonialism to me since I was the only Caucasian in the crowd and obviously for that reason singled out and given this special consideration.

After I was comfortably seated, the stewardess signaled to the others standing near the plane. With one big rush, everyone charged forward, struggling to be first up the steps and onto the plane even though they too had assigned seats. Then I felt glad for my preferential treatment. In retrospect, I see it was another one of those times when I was protected, even though I was unaware of the bedlam that was about to take place.

I fastened my seat belt, sat back, relaxed, and smiled to myself, realizing I was literally halfway around the world from home and finally on the last leg of my journey. Back then, Danang was more commonly known as Tourane, its French name. I praised God for His bountiful supply that had made it possible for me to have finally arrived.

Here, though, I must make it plain that life for me did not become just one big, bountiful supply after another. There was always enough, but I recall one time when I badly needed money to buy stamps to send letters to my many prayer partners. Sometimes I had to wait for a need to be supplied. God always answers prayer either with "yes," "no," "wait," or "not that but something better."

I also must add that through the many years that lay ahead when there was a special and sometimes urgent need, the Lord always came through. Sometimes it would be through unexpected gifts, interesting provision, or special protection, always proving His presence and loving care. There were even times, the provision was on its way before we even knew there was going to be a need. This shouldn't have surprised me, as that was the promise that Jesus gave to us.

> But when you pray, do not use vain repetitions as the heathen do. For they think that they will be heard for their many words. Therefore, do not be like them. For

> your Father knows the things you have need of before you ask Him.
>
> —Matthew 6:7-8

In the months ahead, I sometimes experienced concern and even anxiety. I wondered about my safety and how certain needs would be met. There were two important lessons that I had to learn and relearn: Be content, and be grateful. Again, Scripture is clear:

> And having food and clothing, with these we shall be content.
>
> —1 Timothy 6:8

> Be anxious for nothing, but by prayer and supplication, with thanksgiving, let your requests be made known to God; and the peace of God which passes all understanding, will guard your hearts and minds through Christ Jesus.
>
> —Philippians 4:6-7

Not long after the plane landed, I arrived at WEC Headquarters in Danang. As I settled into my room, I rejoiced at all the ways the Lord had led me thus far. I crawled under the mosquito net, marveled at God's faithfulness in bringing me so far from home, and wondered what tomorrow would bring. Sleep came quickly, and so did the morning.

Vietnamese Language and Cultural Studies

Vietnam is shaped like an elongated S, runs the length of the Indochinese peninsula, and is roughly the size of Italy. China is to the north, Laos and Cambodia to the west, and the South China Sea to the east. When I arrived in June of 1959, it consisted of two nations, South Vietnam and North Vietnam. South Vietnam was a republic, and North Vietnam had a Communist government. The country was divided into two almost equal parts by the Geneva Conference peace settlement in 1954 after the French Indochina War, and the Communists received control of North Vietnam. Danang, my new home, was in South Vietnam about one hundred miles south of the seventeenth parallel DMV (demilitarized zone), the dividing line between the two countries.

Danang is a major port city on the coast of the South China Sea at the mouth of the Han River. When I arrived, Danang seemed like a nice, quiet town. Some of its streets were lined with beautiful, bright red-orange Flame Trees. There were very few cars but many bikes and motorcycles on the roads and some pedicabs, called "*cyclos*" (pronounced see-cloes), three-wheeled vehicles with one wheel in the back, a bicycle seat, pedals, and handle bars for the operator, and in the front, a padded bench between the two wheels for a couple of passengers. It gave excellent door-to-door service, rain or shine.

The *au dai* was the elegant traditional dress worn by Vietnamese women, an indication of social standing for those who worked as shop assistants or who had a higher social status. Long, wide-legged white or black pants were worn under a high-necked, long-sleeved, fitted tunic with slits along each side up to the waist. Lower class workers typically wore loose cotton tops and baggy dark pants. The conical hats were very common and used to shade their faces from the sun. The men dressed in dark, cotton loose pants or shorts and simple dark shirts. Many men wore Western-style clothing. More formally, they wore Western suits, shirts, and ties.

Understanding the religion of the Vietnamese people was extremely difficult. I was under the assumption that it was Buddhism. However, although there were images of Buddha around, I soon learned Buddhism is not the underlying religion of the people. Their religion is not centralized in some building or temple but is ancestor worship, using altars in their homes and shops. Basically, it is a form of animism in which they highly venerate their dead ancestors and believe they receive spiritual power and wisdom from them. There are rituals of worship and various items placed on family altars to honor the dead. Astrology, numerology, and sorcery, I learned, are often considered essential and held in high trust. It was not uncommon to hear of someone going to the *Thay Boi* (fortune-teller) to hopefully ease his or her mind about future events—for a price, of course.

The Worldwide Evangelization for Christ (WEC International) had just entered South Vietnam in 1958 with the intent of bringing the gospel to the many unreached ethnic minority groups. Joan Burridge, a nurse from England, was the one other new WEC missionary who had arrived a couple of days before me. She and I lived together in a large house on the missionary property that consisted of the house of the American field leader and his wife, another house, the house of a Vietnamese pastor and his family, classrooms, and living quarters for Vietnamese Bible school students. Eventually, a chapel was built. Soon other new missionaries would join us from America, Australia, England, Germany, and Switzerland.

A few days after my arrival, I began studying Vietnamese, the official language of both the North and the South. All the basic words have only one syllable, and some words are formed by combining two one-syllable words. I was glad that in the 1600s a modern form of the language had been developed using the Latin ABC alphabet. Before that, Vietnamese used Chinese characters with some additional markings.

In the modern written form of Vietnamese, diacritics are used to indicate different vowel sounds and additional markings are used to indicate the six different tones. The meaning of any basic word depends on which one of the different tones is used. This means that most one-syllable words could be said six different ways with six different meanings, depending on which tone was used, so it was important to be accurate about tones as I soon discovered. Once when our cook was about to go to the market, I reminded her to get a cucumber. She returned with a coconut! Whoops! I had used the right word but the wrong tone.

Vietnam has a tropical climate. Danang is very hot and extremely humid, especially during the months of June, July, and August. We did not have fans in those early days, so after some restless nights, I decided in desperation to leave the double veranda door to my large, windowless bedroom open, which would also cool Joan's room if we opened the door between our adjoining bedrooms. She always kept her flashlight (or torch, as she called it) under her pillow. That night, an intruder tried unsuccessfully to steal it by coming in through my door, going into her room, and reaching his arm under her mosquito net. She woke up and started screaming, "My torch! My torch!" I woke up with a start just as a person ran past my bed and out the veranda door. Somewhat shaken, we closed the door.

Joan had a window in her room (no windows had glass), and she decided to keep the shutters open, as there were bars on the window that we thought would surely keep us safe. Soon after that decision she again woke me up, this time screaming, "My counterpane! My counterpane!" I did not have a clue as to what she was yelling about. All that came to mind was that she was saying something about her windowpane, which made no sense to me at all. Later, I learned she was talking about her bedspread! A man had reached a pole in through the bars of her window and had snagged her bedspread that she had folded over a chair. She awoke just as it got midair and started yelling, scaring the would-be thief away. After that incident, we decided to just endure the hot, humid, airless nights. The shutters were never left open again, and my bedroom door was always shut and securely locked. Lesson learned: Don't invite trouble.

Language study was of supreme importance to me. Every day, five days a week, the eldest son of the Vietnamese director of our mission tutored me for two hours. At first, we used a small language book prepared by a veteran missionary from another mission. We sat across from each other at a table on the wide, covered veranda just outside the double doors of my bedroom. Day after day, I had to learn to pronounce strange, new vowel sounds, learn to use implosive and explosive consonants, study different grammatical patterns, and then there were those essential, ever-present tones. I moved my eyebrows up and down, trying to match them with the right tones, which was tiring and unsuccessful. I then decided to use my index finger (on my lap under the table) in the appropriate direction to match the tones: level, up, down, smooth up down, jerky up down, and hit bottom. This worked well for me.

Tri, my tutor, was faithful in correcting my pronunciation, and later, when I could understand more, he sometimes pushed the books aside and just talked about different, interesting things, which forced me to listen not just to his language but also about happenings in his life. This was an excellent opportunity for me to become somewhat aware of his worldview. Any word or words I did not understand, he wrote down, and I looked up the meanings in my Vietnamese/English dictionary. Sometimes I just guessed.

Once I tried to get Tri to give me a spelling test to see if I could hear the tones and write the words accurately. He was incredulous, as to him that would be a ridiculous waste of time because he was convinced that once the word was spoken, the tones would be obvious. Right? Of course! So much for spelling tests! They never happened.

Once when Tri was sharing an experience with me, I expressed my utter amazement that the people in his story responded exactly as I would have in the same situation. Tri quietly said, "Maybe that's because we are people too."

One day, he told me that he was walking along the road and felt fine, but when he stepped over a hole, he suddenly got the flu. It was caused, he explained to me, by stepping over the hole. I protested and related to him that once I was standing on a corner in downtown Minneapolis waiting for a bus, and I felt fine. Then, just as I saw the bus in the distance, I knew I was coming down with the flu. We went back and forth, but I never did convince him that his stepping over that hole did not give him his flu, nor did the approach of my bus give me my flu.

I soon learned that the Vietnamese people blamed a lot of sickness on what they called "poison wind" *(gio doc)*, which they were convinced caused malaria. I shook my head in disbelief until I realized that the English word for malaria is *mal-air* (bad air)!

A lot of things came up in language study that consisted of more than just learning to speak Vietnamese. Having these discussions, though still in simple Vietnamese with my tutor, was truly exhilarating. In between classes, I listened to endless language tapes and tried to mimic the sounds, remember the words, and repeat the sentences.

When I finally gained some accuracy, I could literally read everything in Vietnamese because it is consistently written phonemically. But there was still the immense task of learning what the words meant. That, of course, took hours, days, weeks, and months in the never-ending task of growing my vocabulary. I made a card catalogue with hundreds of words. Each card listed six words in Vietnamese, each with a different tone marker. Next to each of these words, I listed six English-equivalent words. Although it was a tedious project, it was tremendously helpful. Words, words, and more words! Sometimes I felt like screaming—in English, of course! At times, the whole effort seemed hopeless. I was well aware that my success in learning Vietnamese required dogged determination. I was determined, but it was sometimes quite wearisome.

To learn a language, you must use it. Since we had Vietnamese Bible school students living on the same property, I had endless opportunities to talk and listen. Here I must say that I always found the Vietnamese people gracious, helpful, and patient in my efforts to converse with them.

Joan and I often walked the short distance to town to shop on Saturdays, which was another valuable language-learning opportunity, not to mention firsthand exposure to a different culture. I loved getting out and talking with the people in the open market. I learned how to barter, their common practice, when buying things. It was a strange and somewhat time-consuming experience, but eventually, I learned to walk away from whatever I wanted to buy when necessary, hoping to be called back by the seller and be quoted a fair price. I also learned that it was important to have settled in my mind that I could live without the item, just in case I didn't get called back. It was usually wise to ask a Vietnamese friend the going price of things before shopping. That way, I could barter without cheating them and also without being cheated myself. I didn't mind paying more but not twice the price, which was almost always their first quote to foreigners.

All of our church meetings were of course in Vietnamese, which was tremendously helpful language exposure. Since I was familiar with church vocabulary, I soon understood large sections of sermons. But understanding a lot of what was said was one thing, but saying it myself was quite another. The day came when I was required to give a short testimony in Vietnamese, which was quite easy, as I could read it. The first time I prayed spontaneously out loud in Vietnamese, though, I was pretty nervous.

The language barrier is real. Like trying to get to the top of a high mountain, it takes a lot of determination and concentrated effort to keep climbing. Sometimes, just when I was feeling good about my progress, I would come to another plateau. Suddenly, I didn't understand large portions of what was said. I felt I couldn't learn another word and should just be satisfied with how far I had come, but this was not acceptable, of course. There was still a lot of climbing to do. I realized I had learned a lot, but that was nothing compared to what I still had to learn.

After a year of intensive study, I passed my language test with a score of 93%, but what I had learned had merely gotten me to the lowlands of the never-ending need to learn more. Even so, I enjoyed the Vietnamese people expressing their appreciation when I could finally talk to them in their language and understand much of what they said. This created a bridge between our two cultures and gave me needed incentive to keep learning.

Besides the Vietnamese language, there were many other things to learn and appreciate. During the months of language-learning, I had the privilege, along with the other missionaries, to attend many feasts, some of which were held in rural areas. When we arrived, there would be several rectangular tables joined together in a long row or two, often in the front yard of the church. Down the middle of the tables would be about ten dishes of different kinds of food in a group, with each group duplicated over and over along all the tables. This was so that everyone at each place setting could reach every dish with chopsticks. Now, that was a challenging situation, as no other utensils were provided and probably were not even available. We all soon learned that if you wanted to enjoy the feast, you had to learn to eat with chopsticks. And learn I did, because it was worth it. The food was delicious. Well, most of it. There were one or two items I elected to discreetly avoid, such as cubes of pig fat floating in grease. I noticed that not all of the Vietnamese people partook of every dish either. I never sensed that they would reject me if I did not try all of their food.

As missionaries, we were always treated as honored guests, which was expressed by their making sure that the person sitting near us politely filled our bowls with steaming white rice and then chose one or two delicacies from the middle of the table and added them to our rice. After that we could do our own choosing. These were happy times of conversation with lots of laughter, which often included the cross-cultural misunderstandings we all had experienced. After the meal, the proper thing to do was to leave immediately, as that gave the women and children their opportunity to eat. Except for us lady missionaries, there were only men sitting at the tables at these feasts.

The food in Vietnam is prepared similarly to Chinese food. However, Vietnamese food has a unique character and flavor of its own, mainly because of *nuoc mam* (fish sauce) that they use extensively in their cooking and is served as a dipping sauce with almost everything. *Nuoc mam* is the single most important sauce of

Vietnamese cuisine. It is prepared commercially by layering fish and salt in barrels and allowing them to ferment. The resulting golden liquid is as common as salt and pepper on the Western table. Almost no dish is complete without it and gives Vietnamese food its delicious uniqueness. I was glad that the Vietnamese tend to serve hot peppers separately for which I was thankful. I like a little, but not too much.

When visiting a Vietnamese home, they always served hot jasmine tea in glasses or handle-less cups. The guest's responsibility was to pick up a cup to drink while politely inviting all to partake. When it was time to leave, instead of shaking hands, I learned to close my left hand, cover it with my right, raise them almost up to my chin, bow slightly, and say, "*Toi sin phep ve*" (I beg your permission to leave).

I recall when I really acted foolishly because of a cultural misunderstanding soon after I had arrived in Vietnam. A truck pulled up to the house with my things from the States. The driver and another man carried my trunk and then my crated cedar chest off the truck and into the house. However, they placed my footlocker on the back of a little, thin, haggard, old woman. When I saw that, I strongly protested in the only words I knew in Vietnamese, "*Khong duoc, khong duoc, khong duoc*!" (Not okay, not okay, not okay). The two guys just stood there, clearly not knowing what was not okay, and the lady continued carrying my heavy footlocker on her back into the house. When she came out, they all stared at me, obviously having no clue as to what I was fussing about.

There were also differences between us missionaries from so many different Western countries, which was also a learning experience. Once, I went to a lot of trouble to bake a double-layered, frosted cake, which I thought would be a real treat for everyone at an upcoming missionary get-together. Our kitchen was in a building separate from the house, where cooking was done on a cement platform that had an inbuilt grate over a hole and a place for charcoal underneath. The oven was a small, metal, insulated box with a door that was placed over the grate where coals were burning, and then a metal tray of burning charcoal was placed on top of the oven. It was quite an interesting arrangement and worked beautifully—even for meringue pies!

As for my cake, one missionary decided he didn't like it because it was much too sweet for him. When I asked what his homeland cakes were like, he said they had jam between the layers and whipped cream on top. I wondered how cake with a layer of jam would not be too sweet! Eventually, I came to appreciate and accept many of the different tastes, ideas, practices, and worldviews that delivered me from at least some of my ethnocentrism.

Exposure to different ways of doing things can be frustrating but also educational. There was the time when a crooked chicken-wire basket was upgraded in my value system when we missionaries were going to visit a tribal village, which meant we would have to trek part of the way from the road. I had some covered, plastic food containers (Tupperware) that were given to me before I left home. While we were making preparations for our trip, I told our cook to put the fresh vegetables in those plastic containers, as I just couldn't see carrying them in a misshapen basket made of chicken wire.

The walk into the village was hot and humid, and we finally arrived, feeling quite wilted. Naturally, our formerly crisp, fresh vegetables were not only wilted, they were mushy and spoiled. Then I understood the sensibleness of using a chicken-wire basket, which would have allowed air to flow over the vegetables. I was learning to respect how the locals did things. When I eventually had a small kerosene refrigerator (made in Sweden, imagine that!), the containers were wonderful to use but definitely not for long treks.

A real frustration at first was that the Vietnamese never thought it was important to be on time for meetings—or anything, for that matter. Sometimes meetings were slyly scheduled a half-hour early in an effort to start on time. But the folks soon realized that this was what was happening, so nothing changed.

They have a saying: "Time is like rubber. It stretches." So now I had a choice: Either go with the flow, or get an ulcer. Some things just don't change.

Little did I realize that something enormous and unexpected *was* going to change.

The Australian

On the afternoon of September 4, 1959, three months after I had arrived in Vietnam, I was busy with language study in my room in the house for single-lady missionaries. Our cook arrived at the doorway (we kept our doors and windows open during the day, especially for the welcome afternoon breeze) and informed me that everyone was being called over to the field leader's house to welcome the newly arrived missionary.

We missionaries all gathered around Gordon and Laura Smith's dining room table, enjoying refreshments and finally getting to meet the long-expected Oliver Trebilco from Australia. He had curly blond hair, gentle blue eyes, and held Laura's china teacup like a gentleman. When he spoke, there was no mistaking that he was Australian-born. It was also obvious that he was very joyful and grateful to the Lord for finally being in our midst. It was good to meet him and for him to meet all of us.

As we got up to leave, I was surprised when he said he had something for me. He went into the Smith's guest bedroom and returned with an envelope. While enroute to Vietnam, his ship had stopped in Bangkok, Thailand, where he met a missionary couple who happened to be friends of mine from my US WEC headquarter days. In the envelope was a note from them. Now that certainly was special delivery.

When I walked the short distance back to my house, waiting for me on the veranda with a wide grin on her face was our capable, sometimes outspoken, Vietnamese cook. She had it all figured out that I was going to marry Oliver! *Now wait a minute here!* I thought to myself. I had pretty much given up any idea of marriage. True, I had hoped that God would provide me with a husband who loved Jesus Christ, loved His Word, and demonstrated this in his life. However, hadn't I knelt in the living room of my dorm in Bible school one evening only a couple of years ago, declaring my willingness to be single? Besides, no fellow I knew in either Bible school or at WEC was heading for Asia. When I stepped onto that ship in San Francisco, California, and sailed to Vietnam, I really believed that marriage was not in the plan for me. I shook my head vigorously at the cook and waved her off. But she was still grinning.

Every weekday morning after breakfast, we missionaries gathered together to pray. These were times of sharing and excellent opportunities to get to know each other. Oliver's story was that he was brought up on a dairy farm in a place called Chinchilla, Queensland, Australia. He accepted Christ as his Savior at the age of fourteen, having been sent to Sunday school by his Christian mother. Right at that time, his parents and one of his sisters moved to Victoria, Australia, but Oliver stayed back and worked on the family farm, which now belonged to his brother Ted and his wife, Muriel, who were Christians. There were also two Christian sisters with their families living nearby, Everal Druce and Marjorie Valler.

At the age of seventeen, Oliver felt God was calling him to be a missionary. Kneeling in prayer by his bed, he began to realize that "Go ye into all the world" (Matthew 28:19) was meant for him. However, he was not willing and gave God what he thought were excellent excuses. He was convinced he could never learn a foreign language, and felt he was better suited to be a farmer, but he would certainly support others to go and would pray for them faithfully. God, however, would have none of it, which created a severe problem: Oliver found he could no longer confess his love for Jesus nor address Him as Lord. Two verses of Scripture troubled him, "If you love Me, keep My commandments" (John 14:15) and "If anyone loves Me, he will keep My word" (John 14:23). A short time later, while attending a WEC missionary conference, he was challenged and finally submitted to what he knew to be true—that God's call upon his life was to be a career missionary.

The next step for him at the age of nineteen was Bible school. He did two years at Strathfield Missionary Training Institute in Sydney and one year at WEC Bible College in Launceston, Tasmania. During those years, he had opportunities to visit his parents in Melbourne. When it looked like he was actually going to become a missionary, his father, a rather stern man who professed to be an atheist, became quite upset. So he promised to build Oliver a house and give him his farm, where he could get married and settle down, if he would stay at home and forget this preposterous idea of becoming a missionary. Oliver, wanting to pray about it, tactfully told his father he would think about it.

When it was time for Oliver to return to Bible school, his father wanted to know his decision about his offer. Oliver suggested that he first finish Bible school, and then he could give him his answer. His father took that to be as good as a no and said, "Son, if that is your decision, you will never see my face again!"

In August 1959, Oliver's dad was in the hospital recovering from a minor stroke. Oliver was on his way to board the ship to leave Australia for the mission field, but his dad still refused to see him. Oliver went on his way tearfully yet comforted by the assurance that God had called him to Vietnam. Now he had arrived and would begin language study.

It was not all work and no play for us language students. Sometimes we did some interesting and even fun things together. We went on trips to the interior to visit tribal villages, had late-afternoon picnics on the beautiful white beach of the South China Sea, had birthday parties, celebrated Independence Day at the American Consulate, celebrated Bastille Day at the French Consulate, celebrated the Vietnamese lunar new year, went on sightseeing trips, were invited to Vietnamese weddings and feasts, and of course we met together for our prayer time each weekday morning.

Eventually, there were fourteen new missionaries, which meant different living arrangements. I was asked to move in with a newly arrived missionary, in a section of a different house on the same property, consisting of an office, a large bedroom with shuttered, barred, glassless windows that faced the street, and a bathroom with a flush toilet and shower.

Language and cultural studies continued but with a certain amount of what one might call unrest. As the months went by, thoughts of someone interrupted my studies. And that someone was Oliver Trebilco. I tried hard to not think of him and made no indication of my feelings. But complicating my resolve, every night as I crawled into bed, the voice of a man passing by my window on the street, calling out an invitation to buy his special nighttime snack, filled my ears. To me, the three, clipped syllables he kept yelling over and over sounded like "tray-bee-coe," which is how the Vietnamese pronounce Trebilco. Later, I learned that the "delicacy" being hawked was a boiled duck egg in embryo form. This treat included all the visible veins, heart, organs, beginning feathers, eyeballs, and head with the start of the little beak. I never ate one.

Although Oliver and I were from two different countries and from two different hemispheres, our similar likes, beliefs, and experiences were uncanny. Of special interest to me was when he shared with deep feeling the time he struggled in his Christian walk, which lasted until he yielded to Christ's lordship and experienced wonderful liberation from self and a new ability to be joyfully obedient. His response to God's call upon his life was certainly proof of that.

He had been immensely blessed by the early writings of Norman Grubb, who was the International Secretary of WEC at that time. He spoke fondly of the leaders at the headquarters in Sydney and Brisbane, Australia, as well as missionaries he had met in the WEC Bible school he had attended in Tasmania, whose lives and ministry had blessed him abundantly.

We had both come to Vietnam with the same passionate goal of bringing the gospel of Jesus Christ to unreached tribal people. We both credited Gordon and Laura Smith, our dynamic, experienced field leaders, for inspiring us and making us aware of these neglected people. There were over thirty such groups that we were aware of in South Vietnam at that time.

On Sunday evening, June 26, 1960, Oliver as usual drove us single girls in our mission Volkswagen van back to our house from the downtown house, where meetings were held. The single fellows lived there along with a married couple, Stan and Ginny Smith. When we arrived at the single-girls house Oliver said he wanted to talk to me. When we were alone he said, "You know how I feel about you, Joyce." I guess I did know since he presumed I did! What a missed opportunity! I should have responded by saying, "No, I don't know. You tell me. How do you feel about me?" Well, yes, I did know, because the feeling was mutual.

We were officially engaged on Saturday night, September 10, 1960. He proposed to me on the beach of the South China Sea with a gorgeous harvest moon hovering over the water. The song, "When I Fall In Love," which I first heard sung by Doris Day before I went to Bible school, captures that moment, especially the words: "When I fall in love, it will be forever . . . When I fall in love it will be completely . . . And the moment I feel that you feel that way too, is when I fall in love with you."

None of the missionaries or nationals were surprised when they learned of our engagement, and especially not my cook! I was abundantly blessed by the note I received from my friend Ruth Billman, an older missionary whom I admired. She was thrilled and warmly congratulated us and ended with the comment, "Every pan has its perfect lid."

It was quite a project sharing the wonderful news of our engagement with family and friends in faraway Australia and America. We wrote a joint letter, sent it to everyone, and were thankful for the letters and cards of congratulations we received from so many people. We had engagement photos taken at a shop in Danang to send to our families, as it would be some years before we would meet each other's relatives.

We were married by our field leader, the Reverend Gordon Smith, on January 14, 1961, under a floral-decorated arbor in the chapel on our mission property. We were surprised and honored that the consul and his wife from the American Consulate in Hue attended. I think perhaps their interest was because they too were from Minnesota. At any rate, one day while getting some paperwork done at the consulate, we were invited to have dinner with them there. It was all quite elegant, being served lobster by white-gloved waiters. It was also an honor to repeat our vows to each other before God in the presence of Vietnamese pastors, Vietnamese and tribal Bible school students, fellow missionaries, missionaries from other organizations and some neighbors.

Back home in America, I think my mom must have heaved a sigh of relief. Many years later, when we were returning to Vietnam from our first furlough, she said to me, "I'm so glad you have Oliver." Me too!

Family and friends sent gifts that got through customs easily and surprised everyone. We were also grateful for all the gifts we received from our Vietnamese friends and fellow missionaries. The wedding reception was held in Gordon and Laura Smith's house. All the missionaries helped by providing tasty sandwich canapés, which went well with Laura's hand-cranked, ice-cream-laced fruit punch, salted peanuts, and a beautiful homemade, square, tiered wedding cake. The living and dining rooms were decorated with accordion-fold white bells that my mom sent and white satin bows. Mom also sent lace fabric, which nicely enhanced the locally purchased silky material of my wedding dress. Vietnamese seamstresses are fantastic. They can make anything accurately by looking at a picture of the desired item. My seamstress did a wonderful job. Then to make everything perfect, my brother Phil airmailed me a sparkly, white tiara, which I thought was wonderfully extravagant of him.

After we returned to Danang from our ten-day honeymoon in the tourist town of Dalat, Oliver, by now quite fluent in Vietnamese, was asked to preach at the Sunday morning service. In his opening sentence, he meant to say, "We two, my wife and I, are happy to be here," but instead he mistakenly said, "My two wives are happy to be here this morning." Oh, how everyone laughed, and all eyes were turned on me.

Soon after our marriage, Oliver's then-unmarried sister Doreen came to visit us from Australia. That was special and made more so because she was still with us when we moved to our tribal assignment area.

Grandma Johnson

Mom and her Seven

High school graduation

Myself and Delores

Fort Washington, Pennsylvania

The Castle"
US WEC. HQ Fort Washington, PA

Assigned jobs

Engaged

January 14, 1961

Tribal Assignment

The Vietnamese, or lowland coastal people, are the majority ethnic group of Vietnam. What is not common knowledge is that Vietnam has over fifty (by some accounts, over one hundred) different ethnic minority groups, including those in the North and in the South. They are known as hill people, or *montagnard* (French for "mountain people"). Vietnamese is not the mother tongue of any of these groups, whose language differs from one another. These people do not look Oriental but Polynesian.

After we were married, we were assigned to work among the Hrey (rhymes with *pray*) tribal group at Bato in the province of Quang Ngai with a population of approximately 150,000, one of the two areas where there were large Hrey populations. We drove south from Danang about one hundred miles on the country's one and only main highway to our turnoff at Mo Duc, and from there, we drove west on a winding mountain road about twenty miles inland. We arrived at our mission property that was a mile from the Vietnamese town of Bato. Around and beyond the Vietnamese settlement at Bato were eighteen resettled Hrey tribal villages.

For eight months, while our simple cement-block house was being built, we lived in a small house trailer parked on the mission property. The tropical heat was intense, so Oliver had some men build a thatch roof above the trailer that extended over an area in front of the door, creating a shady place that served as our study. We spent many hours every day there with a Hrey man who came from Ba Heep, his nearby village, to teach us his language. There were no such things as Hrey books, dictionaries, or grammars, but thankfully, we did have an excellent six-page carbon copy of a preliminary analysis of Hrey by an expert missionary linguist, Richard Phillips, which included the alphabet plus a glossary of vowel/consonant combinations and important explanations of the unique features of the Hrey language. This was invaluable in our new language-learning challenge.

By this time, we had both discovered how much we loved language learning, and we were highly motivated. Our purpose for being in Vietnam was to bring the gospel of Christ to tribal people, specifically, as it turned out, to the Hrey people. So now the Hrey were our people, and we longed to be able to communicate with them. Without a language book, we wondered where to start on yet another language barrier. With nouns, of course. So we began pointing to things, eliciting words, and recording the results in notebooks. This was the beginning of our Hrey dictionary, which became a long, ongoing, and seemingly never-ending project. Although tedious, it was also exciting as we moved forward in being able to gradually speak and understand Hrey.

The verbs were trickier to elicit, as we soon found out. In those early days, we tried to find Hrey verbs by demonstration. For instance, we might try to find the word for *drop* by dropping something. Hopefully, we got the right word, but then we might discover later that we had gotten the word for what we had dropped or where it had landed. Of course, the more we learned, the easier it became to elicit words, even verbs, and soon we were able to make whole sentences.

Hrey is not a tonal language, but it does have other unique features like glottal stops and lots of them. Sometimes we use glottal stops in English. We do so when we close off our glottis to exclaim, "Oh-oh!" Some Hrey words have glottal stops at the beginning of words, and at the end. Some words have glottal stops in the middle. Besides glottal stops, Hrey also has breath vowels and non-breath vowels. Two words would sound exactly the same to us at first, but then we would learn that their meanings became different when one was spoken normally and the other was spoken while exhaling. Once we mastered that difference, we then had to learn to say some words with a glottal stop consonant at the beginning, a breath vowel in the middle, and a glottal stop consonant at the end. The difficult part was starting a word with a glottal stop and remembering to exhale for the vowel, but then remembering to stop abruptly for the final glottal consonant at the end of the word. Eventually, after much practice, it was no longer a problem.

Besides learning proper pronunciation, we had to learn many very exacting words. For instance, there were explicit words for "to carry," depending on whether something was carried on head, on back, on shoulder, on pole, in arms, in hand, in both hands, suspended from hand, under arm, in belt, between two people, or between several people.

We not only had to learn the Hrey language but also it was essential that we learn about their customs, their culture, and their worldview. The best way to do that was to visit the people. It was so interesting to arrive at a village with their picturesque houses built on stilts three feet off the ground. The space under the floor of the houses was for pigs and chickens. The walls and floors of the houses were made of split bamboo. Each house was topped with a thatch grass roof. The apex of the front and back of the slanted roofs had thatch-wrapped bamboo poles that protruded to form a V-shape in the middle, which was meant to represent buffalo horns. This style is unique to the Hrey tribe.

The houses in the villages near us were in neat rows as required by the government when they had been resettled from the jungle to an area close to the road. They had willingly relocated because of anti-government guerilla activity in their former jungle village locations. Back there, Hrey houses would be somewhat helter-skelter in arrangement. Also back in their original villages, the houses were very long and could accommodate several families. Each house had one long room with a prepared three-stone fire pit for each family. They slept on straw mats in a circle with their feet toward the pit, which provided welcome warmth in the colder months.

The relocated villages near us had extremely small houses that were barely large enough for one family. There were sturdy, hand-woven baskets placed around the walls that held their supply of rice. Also, there was usually one or more large ceramic rice alcohol jars.

Sections of hollow bamboo tubes that were used as handy pitcher-like water containers leaned against the wall. There was no furniture except a table or shelf, which hung from the rafters, and there were bowls on the hanging tables. The Hrey used their hands to eat.

At both ends of the house was a veranda over which the roof extended and which served as their sitting rooms, where we did a lot of language learning. A notched log was attached to each veranda. When we arrived at a house, we would stand on the ground at the bottom of the notched log and ask, "Is it permissible to come up?" This would bring the response, "It is not forbidden." A grass mat was then brought out of the house and placed on the veranda for us to sit on. Up we would climb, rather clumsily. I always admired the women who could walk up a notched log barefooted, standing straight as a model while carrying a baby on their backs and maybe something on their heads!

The men wore hand-woven loincloths or loose shorts and dark shirts. The women wore dark cotton tops, numerous multicolored bead necklaces, and long, navy hand-woven skirts with red woven trim at the hem. When a woman had a nursing baby, for convenience sake she usually did not bother to wear a blouse. The baby was carried in the ever-handy baby-carrying cloth.

If we thought the Vietnamese were pleased when we could speak Vietnamese, the Hrey were absolutely amazed and thrilled. It soon gave us an instant "in" with them. Their first reaction was disbelief. Then, in utter shock, they would excitedly tell each other that we were speaking their language just in case someone did not realize it.

We found these people to be fun-loving on the one hand and deadly serious on the other. Some of their misguided beliefs produced cruel practices. They are animists, meaning their entire concentration is on appeasing powerful evil spirits. There was no sense of worship, only appeasement.

They told us they were aware that there was a great and good spirit somewhere in the sky, but he did not bother them, and they never asked or even thought of asking for his help. It was the evil spirits—spirits of water, forest, mountains, rivers, ground, etc.—that caused sickness, death, failed crops, and all trouble. They feared these evil spirits and tried endlessly to appease them. One family we knew had a small child who was sick. The parents offered many sacrifices to demons—a chicken, a pig, and then a water buffalo—always after the necessary sorcerer's incantations. Finally, the sorcerer told them to dismantle and reconstruct their house a few yards away to avoid the evil spirits in the ground under their original location. Their fear of demons, a cruel bondage, had left them without needed help and caused them to be further impoverished.

There was a tall pole in the center of every village for sacrificing water buffalo. The buffalo would be tied to it so they could spear it to death and use its blood to ward off evil spirits. This was especially necessary if someone had died, as they did not want the dead person to somehow haunt and harm them or, as they would say, "bite" them. It was sad and shocking that they believed that mankind began by a relationship between a woman and a dog. It was certainly a special experience for us to tell them the true story of creation and of how God created man in His own image. They said they liked that story much better. We felt this upgraded their thinking and their sense of personal value. The truth sets people free.

They also firmly believed that the rainbow was demon activity coming up from the ground. No pot of gold there! When I put a white piece of paper and a glass of water on whichever of their verandas had rays of sunlight shining on it, they were quite impressed with the rainbow the sun's rays produced on the paper as it passed through the prism of the glass of water. It wasn't exactly a Power Point presentation, but it certainly added credence to the truth of the story of Noah and the rainbow of God's promise. It also generated a lot of interesting discussion. God gave us the rainbow, and it is a good, beautiful thing, not something evil.

One of their beliefs led to an unbelievably cruel practice. They firmly believed that your six upper teeth made you look like a dog and that you could never marry unless you had them cut off. Teenagers would taunt each other about being cowards if they did not have their teeth cut off. The peer pressure was incredible. We were saddened when teenaged boys and girls would beg for and submit willingly to this awful practice. A man in the village would have the youth lay on a grass mat. He would then kneel on the ground and with the youth's head between his legs, put a dirty rag in the mouth of the patient, and systematically cut the six upper front teeth off at the gums with a hacksaw blade and a file. We never witnessed this but learned it took about one and a half hours as, one by one, the teeth were cut through and removed. Of course, blood flowed, but water was kept handy to wash the area so the man could see better and keep cutting. The young person might shed a tear or two during the ordeal, but mostly it was an incredible demonstration of unbelievable stoicism. The result was a horribly swollen face with the youth not being able to eat for days. Some kind of black, tar-like sap was applied to the gums. Eventually, the gums would heal, but the beautiful, white teeth were gone. The adults in our villages all had toothless smiles! We missed seeing the brilliant white smiles of our young people after their ordeals. There was fear behind this practice because of their belief that if they had those ugly, horrible teeth, the spirits of spouses would bite each other.

Many of the Hrey people were often drunk. They made rice alcohol in large jars and used long reeds for straws. Several people would sit around a jar, sucking on straws and looking groggy-eyed and red-faced. We saw them offer drinking straws to children as young as four. There were those who used a part of their rice crops for making rice alcohol, which left their families without an adequate supply of food.

Sometimes Hrey men would get together and beat on brass gongs with cloth covered fists creating a dull, monotonous sound for several hours late into the night.

We lived at Bato for three and a half years. During that time, we became quite fluent in the language and began translating the gospel of Mark into Hrey. Since the Hrey people could not read, we began developing a literacy program with Hrey primers to teach them.

I will never forget trying to teach Weang to read. He was a young fellow who had come to Bato from a heavily controlled Communist mountain area. He got tired of their demands to supply them with food, and when they came to get one of his pigs, he objected. He was badly beaten. One day, he and his young wife ran away and arrived at Ba Heep, her parent's village. After hearing their story and seeing the still unhealed welts on his legs, we decided to hire him to do some yard work to help him in his situation. He was about twenty-five, had jet black, curly hair, maintained a lively twinkle in his eyes, and had a wide, toothless grin. How we loved Weang.

Since he worked in our yard and was so available, I began to have him come into the house so I could try to teach him to read Hrey. Our first trial Primer I only had a few pages. After Weang finished it, I decided I should give him a test before going on to Primer II. I chose a word at random and pointed to it. The word was *fish,* which in Hrey is *ca* (in English, this would read *ka,* rhyming with pa). He looked at the word I was pointing to but could not read it. But when he started at the top of the page, he could read *ca* when he arrived at it. How could that be? He had just read the whole book to me! I was disappointed when I realized that he had not been reading at all but had memorized the sentences. I couldn't believe it!

I tried again. Very slowly, I ran my finger along *ca* and slowly pronounced the consonant *c* and then the vowel *a.* Then running my finger quickly along both letters I said, "Ca." But he still didn't get it. At that point, I became exasperated and tapped my temple, "Weang, use your brain, think, think, think!" He looked at me in total amazement and informed me that they do not think with their brains! By now, I was absolutely confounded. "With what then do you think?" It was his turn to look at me in shocked astonishment and calmly answered, "In our hearts, of course." So I asked, "What does your brain do then?" He replied, "It just sits there!" *No kidding!* I thought to myself. Later, I remembered a verse in Scripture: "For as he thinks in his heart, so is he" (Proverbs 23:7a).

Until I had tried to teach Weang to read, he had never held a book in his life, and like all the tribal people, he held a book almost reverently. However, Weang, unlike the other younger people we were teaching, had no incentive to learn to read. A few years later, we would need young men and women to be part of our teacher-training classes. But first they would have to learn to read their own language.

Oliver went to many villages to try to enlist young sons and daughters. He would gather the tribal elders in a village together on someone's veranda, where they would listen to his appeal politely. Their response was equivalent to asking why in the world their young people needed to learn to read. How would it help them do a better job tending the water buffalos, farming, and doing chores? Oliver would then take one of our books from his bag and begin reading to them some of their folklore stories that we had recorded. They looked at each other in disbelief and listened intently, impressed. Here was a white man who had flown in on a US military helicopter speaking their language and reading in Hrey some of their traditional oral stories from a book! Eventually, we did get some of the young people to come and be trained to teach.

During our time at Bato, it was my great joy to teach young people who came to our house to read and write Hrey. I loved those literacy classes and those dear young people. Later, many of them were killed as soldiers in the war. Oliver also took the opportunity to share the gospel with my group. One of them was a shy young man named Ngiah. He was not pretty to look at since one of his eyes protruded from the socket about an inch. He had been "doctored" by someone as a child with the application of some kind of crushed leaves to his bad eye, and the result was tragic. Ngiah always stayed in the back of the group. My attention was on a handsome teenaged boy who looked like he could help us in our work one day, but little did I know that it was Ngiah who would eventually become our most valued ministry partner.

We took Ngiah to Danang, a new experience for him, so he could have this ugly eye removed by Dr. Herbert Billman, our missionary doctor. Herb kindly ordered a glass eye for Ngiah, which was much appreciated when it arrived. We had to smile when Ngiah noted a uniformed man standing in the middle of an intersection waving his arms around. He didn't know about policemen who directed traffic, so Ngiah thought the guy was crazy.

One evening while still in Danang, as Ngiah was recovering from his eye surgery, I was talking with him, and he began speaking of Jesus. As the conversation progressed, I asked him: "Where is Jesus now, Ngiah?" He replied, "*Haq oi ta manoh au!*" (He is in my heart!). Ngiah had not only attended my literacy classes but also came to our

Hrey language church services to hear Oliver preach. One day when he was returning to his village, a truth became clear to him as he was walking along the road. He told us that he was thinking about what he had heard in church and thought, *The missionary is right. Jesus is the true sacrifice. He bore the penalty for my sin, and His sacrifice was for my salvation, so I do not need to appease demons any longer.*

When we returned to Bato there was now a curfew because of nighttime danger that would have prevented the young people from being able to come for my literacy class. We had a tribal house for Weang and his wife built on the mission property close to our house, so the young people stayed there in safety after my class.

Every Sunday morning, our Vietnamese pastor held a well-attended Vietnamese worship service at the church on the property close to the road, which we always attended. The Vietnamese pastor usually preached, but sometimes Oliver did. After that service, we would hold a service for the Hrey people in their language, with Oliver doing all of the preaching. We were thrilled to see change gradually take place in people as they turned to Christ and were delivered from agonizing, tormenting fear of demons.

I have a special memory of Ngiah's initiative and thoughtfulness. I was getting ready for the children and young people to put on a Christmas program. For the shepherds, I used cloth and ties for their headpieces, and decided their shorts and shirts would suffice for the rest of their outfits. Before the program, Ngiah produced shepherd's crooks that looked authentic, probably from a picture he had seen. He had shaped them after soaking the ends of bamboo in water; how perfect for a Hrey Christmas program!

Rai was an old, thin Hrey man who faithfully attended our services at Bato. Once when Oliver was preaching about Jesus being the Good Shepherd and that we are like sheep, hearing His voice and following Him, old Rai piped up, "I'm one of His sheep. I hear His voice. I follow Him."

Some time after this, Rai's five-year-old grandson died. Rai insisted he have a Christian funeral, which of course, challenged the entrenched, fear-dominated beliefs of the Hrey. They believed that if you did not do right by the dead person, he or she would come back and bite you. A chicken's neck had to be wrung and placed on a tray, as this chicken was the necessary key to get you into some place. They didn't know where, but you could only get in if you had that key.

When someone dies, the body is wrapped in a straw mat and left in the house until all the preparations are made. A buffalo is sacrificed, and the blood is smeared on the doorposts of the house or on a fringed pole placed diagonally beside the doorway. They then did what they called "dividing of the possessions." They chose various parts of the buffalo that they had sacrificed to demons and carried them suspended from a pole to the grave to insure that the person would have a buffalo in the next life. They also took rice, bowls, and other utensils to the grave and left them there for the person to use in the next life. Old Rai was adamant and would not allow any of this for his grandson.

We were touched to be with him as we climbed a hill with the people and buried the child. Rai stayed at the gravesite until every person had left to insure that nothing meant to be used in the next life was left behind, as Rai did not believe any of that any more. He had embraced the gospel with the assurance of a real place called heaven where nothing from this life would be needed. We stayed with him until everyone was gone and then walked back down the hill together. This was the first Hrey Christian funeral in that area.

Rai instructed everyone that his funeral would be a Christian one. We were not there when he died, but we are pretty sure that the people were more afraid of his spirit biting them if they did not follow his wishes to have a Christian funeral as per his instructions rather than follow their heathen traditions.

We were encouraged by such results and kept busy in our various ministries. Our house was finished in September 1961, and for several months after we moved in, everything was peaceful. Suddenly, though, all that changed. The enemy stepped up its efforts to force the people to leave their government-relocated villages and return to the mountains.

Fire! Fire!

We had hoped to live in a Hrey village, but the Vietnamese authorities never granted us permission. "Too dangerous," they insisted. However, our house on the mission property was near the villages, especially one right next door that turned out to be close enough.

Just ten months after we moved to Bato, our somewhat peaceful situation changed drastically. Anti-government insurgents began burning down Hrey villages, threatening the Hrey people with death if they did not return to the mountains. The insurgents wanted this so they could exploit the Hrey as they had done before the government had relocated them. In one village, one hundred and twenty two people finally did return to the mountains. But some, refusing to submit to the terrorists, fled and hid in their fields and spent the night there lest they be captured. The young men and boys especially feared being forced to become soldiers for the enemy. The next night, that whole village was burned to the ground.

The next morning, we visited the village. People stood dazed among ashes that was once their village. It was cold, wet, and miserable. All their homes were heaps of ash with only some charred poles still standing. groups of people were sitting in the mud. We were able to provide them with straw mats to at least keep them off the ground and obtain donated clothing and blankets for them. Children huddled together on those mats, and adults were already starting to put up temporary thatch and bamboo shelters to protect them from the sun and rain. They eventually built new houses, even smaller than before. Once again, they hoped that the government would be able to protect them. Over a period of time, fifteen villages around us were burned down, totaling about one thousand houses.

On January 18, 1962, we were awakened at about 11:00 p.m. by shooting at the back of our house and on the road at the entrance to our mission property. Hearing the whine of the bullets, we shot out of bed. Just as we thought. Another fire! This time, it was right next door—so close that our whole compound was lit up and we could hear the fire crackling. As the village burned, we heard what sounded like a lot of shooting. This seemed to us like a land attack, which would be really serious because this meant that hundreds of men would be pouring in with blazing guns. Actually, it was not gunfire that we heard. Because all Hrey houses were built with bamboo poles, the heat from the fire caused the airtight sections of many burning bamboo poles to explode simultaneously, sounding like machine gun fire.

The pattern for all the village burnings was the same. At about 10:00 p.m., enemy guerrillas would surround a Hrey village, and they would order everyone out of their houses. The people would plead with them for mercy but were told to "get out or get shot." Men with torches would then set the houses on fire. The whole village would instantly be ablaze, and in only a matter of minutes, everything would be burned to ashes. The next day, and until they could build again, was a miserable time for the Hrey without proper shelter.

In one of the nearby villages, a family had gotten into a foxhole they had dug, and to their horror, the burning house fell on them. Sadly, mother and father were badly burned on their faces and their infant son on his scalp. We quickly drove them to the American Special Forces camp at Bato, but their burns were too severe to give them adequate care there. The American military, in just one of their many missions of mercy, flew them by helicopter to the provincial hospital in Quang Ngai where they received much-needed care. Many weeks later, we were so amazed when they returned to visit us. Their healing was wonderful, and their toothless smiles and happy eyes so beautiful to us.

By now, because of the escalating danger, we were thankful that we had not been given permission to live in a village. By September

1963, there were ever-present reminders that the country was trying to rout the enemy. Sometimes we heard the drone of engines, and looking down the road, we saw the long, snake-like line of Vietnamese army trucks making its way toward Bato. There were scores of trucks as far as the eye could see, and still they came. Deep in the night, the stillness was often suddenly shattered by the blast of Howitzer 155mm guns. Sometimes we heard the round going right over our house, making a loud whirring sound as it headed toward the bad guys in the distance. The blasting would continue, shaking our bed. When dawn was barely streaking across the morning sky, several helicopters would arrive and set down at the US Special Forces airstrip. The road and air traffic could continue sometimes for a week or so, as another big operation to clear out the enemy was in progress. Sometimes we saw tanks on the road and later would hear the roar of their guns in the next valley.

Other measures were taken. All able-bodied Vietnamese and Hrey were called upon to help surround the town and the villages with two barbed-wire fences with a mote in between lined with sharpened bamboo sticks. We were now living in a war culture. Everything was changing.

I began to feel fear tugging at my heart, and Scriptures about fear became more than just words to me. One verse seemed especially unique: "Whenever I am afraid, I will trust in You" (Psalm 56:3). Until I had some unnerving experiences, I never once considered the possibility of fear and faith coexisting. We did not know if we would survive, but evidently it was not impossible to trust and fear at the same time.

Despite many uncertainties, we remained at Bato in faith, believing that God had sent us there to do a job. Some Hrey were turning from their demon sacrifices to Christ in those early days. We were learning their language and were teaching some of them to read and write. Then we began to translate the Scriptures into Hrey. But little did we realize the long road that lay ahead of us. We did experience the Lord's presence though, and were aware of His loving care, even when things were difficult and uncertain.

Actually, we got things backward at first. We were so enthusiastic about getting God's Word into the Hrey language that, as soon as we could, we began translating the gospel of Mark. The problem was that not one single person among them could read Hrey. We soon realized the need of a literacy program. It was then that we received the most welcome opportunity to be involved in the Highlander Education Program sponsored by the government. This was a huge project that required developing reading primers, simple science, folklore, math and health books, preparing teacher's manuals, recruiting potential teachers, teaching them to read Hrey, and then training them to teach the prepared materials. This became a time consuming undertaking for us, and made more difficult because of the war. Yet we were motivated by our intense desire that the Hrey people not only have the Bible in their language but also that they could read it. We pressed forward, but there were always those inevitable delays.

Sicknesses

Looking back, I wonder why we didn't contract tuberculosis, as sometimes we were around those who were afflicted with the disease. However, we did not escape malaria and were stricken with it more than once. Having malaria is something like having a severe case of the flu but much worse. The unbearable headache, the inability to keep food down, the aching bones, and the wretched high fever and chills were disheartening, to say the least. No wonder we would declare, "At first I was afraid I was going to die, but then I was afraid I wasn't!"

The prophylactic we were advised to take once a week was a horribly bitter pill that would hopefully prevent us from getting malaria. And it did. We took ours every Sunday, but one week I foolishly forgot. By Wednesday, the onslaught was terrific! Soon my temperature went up to 104 degrees; I had teeth-chattering chills, a pounding headache, and all-over body aches that made me so miserable, especially during the long, long nights, as I could not sleep. Oliver would rub me down with cold water, which helped for awhile. I could neither eat nor take any kind of medicine because nothing would stay down.

On the afternoon of the tenth day, I felt myself sinking to an even lower level of weakness and languishing misery. I asked Oliver to pray for me again and to anoint me with oil as we are instructed to do in Scripture:

> Is anyone among you sick? Let him call for the elders of the church, and let them pray over him, anointing him with oil in the name of the Lord. And the prayer of faith will save the sick, and the Lord will raise him up.
>
> —James 5: 13-14

Oliver rubbed a little of oil on my forehead and prayed. By that evening, I was able to keep down a bowl of soup. I was on my way to recovery, but it took days of gradually regaining my strength before I felt really well once again.

Another time, Oliver got hit with a severe malaria attack. He was so nauseated that he sat on a stool between the bathroom and the bedroom. Once after he had weakly climbed back into bed, he called me into the bedroom and asked me to pray for him and to anoint him with oil. I went into the kitchen, poured a small amount of olive oil in a little dish, and returned to him. I knelt by the bed and, reaching under the mosquito net, rubbed a bit of oil on his forehead and prayed. I had barely said amen when suddenly he sat up, threw both of his arms up in the air (so out of character for him), and began loudly praising the Lord. By now, it was time for our evening meal, and he was able to get right out of bed, sit at the table, and eat a big meal, just like that! He didn't even need the usual long recovery period.

Being obviously touched by the Lord in our sicknesses was very wonderful and encouraging to us, especially since we were living in a somewhat remote area. Perhaps that is the reason we saw those results. It certainly quickened our faith and blessed us abundantly. However, for whatever reason, we have not often experienced such dramatic results since those years at Bato. Yet there is that bubble of joy in my heart when I remember those special incidents. Truly the Lord was with us.

The upside of our getting sick with malaria was that it removed our criticism of some of the tribal people. Sometimes we saw someone sitting on a veranda in the village, doing nothing for days, and assumed that he or she was just lazy. After having malaria, we empathized with those people because we had experienced firsthand malaria's weakening effect and the long road to recovery.

Oliver did a lot of medical work in the villages almost on a daily basis. He had received good training from our experienced missionary doctor and was therefore able to help many sick people. Sometimes there were only minor problems, and just an aspirin would work powerfully on patients because they were not used to taking any kind of medication. There was one interesting thing they told us. When they got medicine from the government clinic, they said it did not work as well as our medicine. Actually, it was the same stuff! Perhaps it was because they referred to our medicine as "The Lord's medicine."

Oliver had learned what medicines to give for the obvious symptoms of malaria, earaches, colds, and diarrhea. Of course, it wasn't perfectly scientific since there were no tests to pinpoint the exact problem, but he certainly brought welcome relief to many suffering, grateful people. Sometimes it was necessary to prescribe a course of pills to be taken over several days, and Oliver always carefully explained the dosage schedule. In a day or two he would return to check on the person and was often dismayed to discover that the patient had taken only one pill and, since he or she did not experience immediate results, decided it was useless to take the remaining pills. The explanation was usually, "I took one, but it didn't work."

Oliver would then have to start all over. First, the packet of unused pills had to be found and brought out from wherever it was tucked, usually somewhere up in the rafters. He would then give his lecture again, explaining the importance of following the medicine-taking schedule every day right down to the last pill. No matter how many times he explained this, they gave the usual Hrey answer for anything they neglected to do, *"Au het au"* (Me, I forgot).

Sometimes we were faced with heartbreaking situations. An old Hrey man came to us from back in the hills with his middle-aged son. They had walked many miles to get to us. The son had a tumor the size of a large grapefruit on the side of his terribly deformed face. Oliver took him to the American Military clinic in Quang Ngai, but they just sorrowfully shook their heads, as it was too late to do anything. He was given some pain-killing tablets, and they returned to their village. I will never forget the look of despair on the old man's face as they turned and walked back to their jungle home. Another time, a Hrey child had picked up a discarded grenade, not realizing it was still live. It blew up, and part of his face was blown off. Once again, it involved a trip to the hospital in Quang Ngai, but sadly, the child died.

Many times, the enemy guerillas would stand on the hill behind our house and shout their anti-government slogans through a megaphone. The tribal people said they had asked them if we had guns. When they replied in the negative, they were told that if we had guns, they would come and kill us. These were local insurgents who would have been aware of who we were and known all about our work and activities.

Every year at Tet, the Vietnamese lunar new year, the insurgents would shout their murderous slogans from the hill behind our house that this was going to be the year they would come and "drink blood." I don't recall having any real concern when hearing this, perhaps because they said that every year and it did not happen. We were in close contact and fellowship with the Vietnamese pastor and his wife who lived on the same property. Since they were not concerned, we did not feel the need to worry about the yearly threats of a ruthless takeover of South Vietnam. However, the year would come when those threats would become a reality, but not yet. Though we were mildly aware of danger, we were able for the most part to carry on with our various ministries, for which we were grateful, realizing it was just the way things were. There were times, though, when the sense of danger was not so mild.

Dangerous Roads

Our winding, twenty-mile mountain road from Bato to the highway was often mined by anti-government forces during the night, which was troublesome. Whenever we had to travel that road, we usually waited for the Vietnamese military to do its routine morning check for mines before we ventured out. We also had a policy to never travel in a military convoy, which would welcome danger. Our Land Rover had a large white cross painted on both side doors with clear words in Vietnamese identifying us as belonging to a Christian mission organization.

Sometimes it seemed in the early days of the war that insurgents considered missionaries to be okay because they felt we were helping the people like they were supposedly doing. However, it was extremely disconcerting when in a different area of Vietnam, Roy and Daphne Spraggett, a WEC missionary couple, and their two-year-old daughter were severely wounded from the blast of a bomb that had been planted in their house. But so far, we were able to continue our ministries of village visitation, teaching, preaching, medical, literacy, and translation as usual.

One day, there was a medical emergency, and Oliver was asked to take a sick Vietnamese woman to the hospital in Quang Ngai. This was not an unusual request since ours was the only non-government Land Rover in the area. But I did not mention to Oliver that I had an uneasy feeling about this particular trip. He left right after breakfast with the sick woman and two of our Vietnamese pastors.

I could not shake a feeling of foreboding. I went into our bedroom, knelt by our bed, and began to pray for their protection. After a few minutes, I was reminded of part of a verse in the Old Testament: "No weapon that is formed against you shall prosper" (Isaiah 54:17a). I prayed that it would be so. Of course, there were no phones or other means to contact each other. That was an incredibly long day for me.

In the evening, Oliver returned with good news that that he did get the sick woman to Quang Ngai where she was being cared for in the hospital. There was more to the story, however.

He told me the day had started out with an uncomfortable feeling that morning when he left. Soon after he was on the road, he was reminded of a Bible verse: "You shall not be afraid of the terror by night, or of the arrow that flies by day" (Psalm 91:5). He wasn't thinking of arrows, of course, but of bullets. Feeling relieved and somewhat safe since it was broad daylight and our vehicle could be identified as ours and not a military vehicle, he drove past rice fields on either side of the road.

Suddenly, he noticed bits of earth being kicked up on the dirt road in front of the Land Rover. His first thought was that they were raindrops, but it was a nice, sunny day. Horrified, he became all too aware that it was a spray of bullets hitting the dirt. He had not heard the sound of the gunfire because of the Land Rover's loud diesel engine. Realizing that he was the target, he accelerated.

Soon he approached the creek where the bridge had some days before been blown up by the guerrillas. This was about five miles from our house. As expected, there were Lambrettas on either side of the creek. These three-wheeled Italian vehicles were a common means of public transportation, and each accommodated up to ten passengers sitting on two benches on each side of the three-wheeler. Sometimes two passengers would occupy the seats on each side of the driver. Because the bridge had been destroyed, passengers coming from or going to Quang Ngai would have to exit a Lambretta on the opposite bank, walk down through the shallow creek and up the other side, and get in the waiting Lambrettas, which would then take them to their desired destination.

Because the Land Rover is a four-wheel drive vehicle, Oliver would ordinarily have driven past the parked Lambrettas, down through the creek, up the other side, and continue on to Quang Ngai. But this time, a sudden strong inward impression said, "Stop!" He hit the brakes. Just then, a spray of bullets from the trees in the distance hit the empty Lambrettas in front of him and to his left. Oliver jumped out and got behind the left rear wheel, while his passengers jumped out and scurried into a rice paddy and lay flat on their stomachs. Soon Vietnamese in black, pajama-like clothing appeared. Oliver was pretty sure he was going to be killed or captured. The black pajama guys, though, turned out to be the Vietnamese Home Guard and repelled the enemy. Everyone piled back into the Land Rover, and Oliver drove down the creek bank, up the other side, and continued on his way to Quang Ngai.

I was shocked to learn he had stopped at the creek on the way home to take pictures of where all this had happened. Later, we learned that our Hrey friends had somehow communicated much displeasure to the enemy, who apologized, saying they didn't realize it was the missionary. The apology was passed onto us. Another time, though, they shot at us as we were a short distance from our house, the opposite way along that same road. Oliver hit the gas peddle, and we roared home.

On another trip to town, we had a different kind of dangerous road experience. We were winding our way from Bato on the narrow road toward the highway. The Vietnamese pastor's wife was sitting next to me in the front seat. Suddenly, looming up in front of us as we came around a bend was a military truck—completely on our side of the road. Crash! My knees hit the metal bar under the glove compartment, making two deep dents in it. My knees were only slightly hurt, which could only have been a miracle, and a muscle in one of my feet had gotten painfully twisted. The pastor's wife was thrown into the windshield, smashing it, but happily, she was not seriously injured. Oliver was unhurt. There was a sudden, eerie silence after the loud crash of metal and glass. We were not only stunned but also apprehensive, as we became concerned, realizing that we might become the target of enemy snipers.

To our utter surprise and relief, a military jeep appeared behind us within seconds, driven by a South Vietnamese army captain, who was a friend of ours. He apologized profusely, realizing that it was the truck driver's fault. He drove us all back to our house at Bato. Oliver made a crutch for me out of a mop handle so I could hobble about on my hurt foot.

Meanwhile, our captain friend had our Land Rover loaded on the back of a military truck to be taken to Quang Ngai to a repair shop. Oliver went with it, returning the next day. It was many days before the Land Rover was fixed and back in service again. We never pressed charges against the other driver. The pastor's wife and I eventually recovered from our injuries.

We knew that the Vietnamese drivers of military trucks were afraid of falling off the unguarded edge of the mountain road, so they tended to hug the mountain even when they were on the side of oncoming vehicles. Also, they were well aware of the possibility of encountering enemy guerrillas firing down at them from the hillside.

Soon after this incident, the same thing almost happened to us again with the same driver in a military truck. This time, he flashed us a quick smile as he barely squeezed past us on his side of the road and went on his way. But we were not smiling!

The Chocolate Candy Bar

In December 1962, we were living in our simple but cozy cement-block house that a Vietnamese contractor built for us at Bato. It was located in a valley surrounded by green, jungle-clad mountains. On the same mission property down close to the road was the church, the house of the Vietnamese pastor and his family, and, next to it, the house of a Hrey family who were involved in our ministry. Some distance from the road and up a slight incline were two missionary houses. One of them was ours, and the other one was empty. We used that for our Hrey church services.

In those days, we had American post office (APO) privileges, which meant that any mail that came from the States came via US military mail. Not only was this a wonderful arrangement and a fantastic privilege for us, it was also helpful for those who sent us mail and packages from the States, as they only had to pay for postage to California. From there, our mail arrived first in Danang and then was flown to the American Special Forces camp, which was a mile down the road from us. When we saw a plane arrive, we waited hopefully to see if a jeep would come racing along the road toward us, bringing our mail.

This particular day there was mail for us, so it was not long before tall, handsome American Captain Rose arrived at our door. I had just started wearing maternity clothes, as our firstborn, Janice, was on the way. Captain Rose, a true gentleman, took one look at me and said, "Congratulations!" We chatted for awhile, I accepted our mail, and he stood to leave. As he put his hands in his jacket pockets, he seemed surprised to discover something in one of them. He took it out. It was an American chocolate candy bar. He held it out to me, asking if I would like to have it. Delighted, I said, "Well, yes, I would!" Unknown to Captain Rose, I had been craving chocolate, real chocolate, for some time. In Vietnam in those days, if you wanted chocolate candy, you had to make fudge, which to me is not real chocolate. This incident was so insignificant, yet it was a special and awesome reminder of God's tender, timely, personal, and loving care for me in that faraway valley.

After Captain Rose left, I opened the candy bar. Real chocolate! I sat down and relished each delicious bite with a smile on my face and a big "Thank You!" in my heart.

Daughters

We welcomed our firstborn, Janice Alane, on May 8, 1963, while we were serving at Bato. Her name means "God's gracious gift." We also gave her a Vietnamese name: *Thanh Hoa* (Blue Flower). In the city of Quang Ngai, the midwife in charge of the maternity ward of the Vietnamese provincial hospital was a capable Canadian. She was quite proud of herself that she had made the nurses remove all the bottled placentas from sight, as some Vietnamese women believed that there was medicinal value in them. At any rate, all was in readiness when I arrived there. She used Janice to demonstrate to new mothers how to give a newborn a bath, which I found quite amusing. My firstborn was already a model!

Missionary friends from a different mission, Jim and Jean Livingston opened their home to us for the time before and after the birth which was a fantastic provision. The morning after Janice was born Jean sent over a breakfast tray. I enjoyed that meal so much, especially the freshly baked buns. I was so hungry.

After we left the hospital we returned to Jim and Jean's house. When we arrived, Jean had a bassinet prepared for our new baby. The mattress was encircled with gorgeous pink double hibiscus blossoms. It was so beautiful and thoughtful. What wonderful friends! We were so happy.

However, at this same time, we found ourselves in a bit of a dilemma. The Land Rover we had use of at Bato belonged to another missionary couple who were on furlough and would need it as soon as they returned to Vietnam, at which time they would be moving to another area. We urgently needed to own our own vehicle. Jim had decided to sell his Land Rover since he and his family were about to go on a one-year furlough. Oliver, so eager to buy it, could hardly sleep for several nights just thinking about it. If we were to buy the Land Rover, we would pay half immediately and the other half while they were on furlough. The problem was that for the first half we needed $500.

A few days after Janice was born, we were in their guest bedroom with our newborn daughter sleeping in a bassinet at the end of the bed. I was resting, and Oliver was kneeling by the bed. Suddenly, he said, "I am not going to ask for the money any more. God knows we need transportation! I am just going to begin praising Him for it!" And that is exactly what he did. His outburst surprised me, as it was not like him.

No one among all our friends and family in America or Australia knew about this need. Then we received a letter from a couple in America whom we had never even met. They explained that they had told some friends of ours that they had some finances they felt they owed to the Lord, and our friends mentioned our name. And in their letter was a check for $500! The perfect timing of its arrival was more wonderful than words can express. Like Oliver had said, God knew we needed transportation. Although we had until May of the following year to finish the payment, we were able to pay in full by November. God had even supplied enough money for us to buy new tires!

We returned to Bato shortly after Janice was born and were now referred to by the Hrey as *Miq Baq E-Jan* (Mother, Father of Janice). *E* is the Hrey marker for female names. Janice had become famous in Hreyland just for being born!

How the Hrey loved her! They had never seen a white baby. They marveled at her blue-grey eyes and her virtually bald head. They thought I had shaved her head and wanted to know why! Their babies are born with lots of thick, black hair. I assured them I had not shaved her head, and that actually I was waiting patiently for her hair to grow. I sometimes attached a bow to the top of her head with tape. I liked the look. It was months before Janice's blond curls finally appeared, which of course made her unique in Hreyland.

It was almost time for our first furlough. We were expecting our second child, so we moved to Danang where we had made arrangements for the event at a small maternity hospital there. Right at that time, the authorities at Bato informed us that they would no longer allow us to return there to live, as it was becoming more and more dangerous. However, Oliver did return one day to get some of our things.

God gave us another daughter, Jeanette Aline, at 3:15 p.m. on August 5, 1964. Her name means "God's Gracious Gift," which she was and still is to us. We also gave her a Vietnamese name. Although August is not quite autumn in Vietnam, we chose the name *Thu Hoa*, (Autumn Flower).

When I saw her for a brief moment soon after she was born, I smiled at her dear little round face and tightly scrunched-up eyes, rejoicing that Janice had a sister. Janice, meanwhile, was well cared for by a faithful servant girl back at the house while Oliver stayed with me.

I had some serious complications that required blood transfusions. The maternity hospital did not have a supply of blood, so Oliver and a nurse carried me on a stretcher from the second floor of the hospital, down the steep steps on the outside of the building, and placed me in the back of our Land Rover. He then drove to the Danang General Hospital, where we soon discovered that they did not have my blood type. Now what would we do? It was already getting toward evening, and there was a firm curfew, especially for foreigners. Oliver nevertheless immediately left the hospital and drove to the Danang Hotel where US military personnel were billeted. He rushed into the hotel, announced our problem in the lobby, and asked if anyone would be willing to give blood. Soon four guys with my blood type piled into the Land Rover, and Oliver drove furiously to the hospital where these men kindly donated their blood to me. Of all the wonderful blessings we had received from American soldiers, this was certainly one of the most special. We were so grateful, and I have so often wished I knew who those men were.

Late that night, I was brought back to the maternity hospital and reunited with my brand-new daughter, Jeanette. Now our family had four members.

In December 1964, we boarded a French ship that took us to Singapore, and from there, we flew to Oliver's family in Australia. This was my first time in Australia, and I experienced a certain amount of culture shock. I had already learned several language differences. What I called a diaper, the Australians called a napkin; what I called a napkin, they called a serviette. A cup of tea was called a cuppa, a cookie was a biscuit, a baking powder biscuit was a scone, and a farm field was a paddock.

Jeanette was four months old when we arrived in Australia. We were met by many family members, and I will never forget one of Oliver's teenaged nieces rushing up to me, pleading, "Can I have a nurse? Can I have a nurse?" To me, that meant she wanted to breast feed Jeanette, to nurse her! Not so. To nurse in Australia, I learned, means to hold. Later, when we got to America, Oliver was telling my mother all about koalas. She was extremely interested and asked if he had actually seen one. He proudly exclaimed, "Oh, yes, and I even nursed one and had my picture taken." My poor mother just stared at him and, we think, blushed a little. Same problem. He was actually telling her that he had held a koala.

We had a wonderful time in Australia where we were invited to tell about our missionary work in many meetings. It was especially my pleasure to meet all of my husband's family. His brother and two sisters lived in Queensland, so we spent a week with each family. I am so thankful for that, as it gave me excellent opportunities to get to know them and to feel like I was part of their family. They were delightful people, very Australian, and always made me feel welcome. They teased me, of course, about what they called my accent. (*My accent?*) Each family was so kind. They took us to many places of interest and made sure I saw a lot of the real Australia. For that I was grateful because I was on the inside of my husband's Australia, not on some Americanized tourist trail.

One evening while staying at Oliver's brother Ted's farm, Ted announced that he was going to take us to see some "roos." He assured me that we would see mobs of kangaroos from the back of his truck when they shined the spotlight on the paddocks, but I didn't believe him. I thought that would be like saying we were going to sight a herd of deer in a Minnesota meadow. It just wouldn't happen. But I was wrong. Sure enough, I saw mobs of kangaroos and watched as they bounced along. What a sight! I was impressed.

Our family also drove to Victoria, Australia, on that our first furlough to stay with Oliver's dad and mom, which was somewhat awkward, as his estranged dad left the house before we got there. But circumstances brought him back. We were grateful for the time we had with them. Our little daughters no doubt made it easier. During this time, on some evenings, Oliver was speaking at several churches in the area. I noticed that his dad kept watching the clock and then at a certain time would start to boil water for tea. I finally realized he was getting ready to serve Oliver a "cuppa" when he got home. This was done quietly. Observing this man trying so hard to appear disinterested so as to keep to his stubborn vow of long ago was interesting.

I had a nice time with Mom Trebilco. She had some interesting scrapbooks of Queen Elizabeth II that she was eager to show me. Sitting with her, looking at the pictures, and listening to her comments was actually quite educational. I wish I had gotten to know her better. One of Oliver's sisters, Doreen, was also there. She was single and lived with them. I had already met Doreen, as she had visited us in Vietnam just after we were married.

The second part of our first furlough we spent in America. We stayed most of the time with my mother and stepfather in Minneapolis. My family also did a lot of things to make our furlough memorable, and it was my joy that they all got to finally meet my husband and our two daughters. We traveled a lot by car to different states for meetings. We took our girls with us. Sometimes we dressed them in either Vietnamese or tribal outfits, and they would sing at meetings. People loved that. We are thankful that our family was blessed with many people who would be praying for us as we returned to Vietnam. More than one person told us that they prayed for us every day. Every day! We were astonished, touched, and blessed by this interest in us.

We were privileged to speak in scores of churches. People were extremely interested in what God was doing in Vietnam. They already had plenty of war news, so sharing about the Hrey and many other tribal groups hearing for the first time and responding to the Good News in Vietnam, was wonderful. Also, although the war was ugly and sad, we were grateful for how it kept the country open for many tribal groups to hear the gospel for the first time, turn to Christ, and become grounded in their faith. We think of it as the other war in Vietnam, the spiritual war.

On this furlough, we studied linguistics for the summer at the University of North Dakota, which was extremely helpful. Janice was with us, but we left Jeanette with my mom in Minneapolis. In August, when Jeanette was turning two, my sister Kathy and Mom brought her to North Dakota to celebrate. When they arrived at our dorm, they sent Jeanette alone ahead of them. When she met me in the hall, she looked up at me and asked, "Are you Mommy?" Talk about a tear-jerking moment!

Although our time at home had been somewhat extended, it still went by quickly. Soon, it was time to return to Vietnam. My mother and stepfather went with us by train to San Francisco. Once again, there were sad good-byes. Oliver, our small daughters, and I boarded the SS *Wilson* of the President Lines. On the deck, we were given rolls of paper streamers. I threw mine down to my mom, and she held her end and I held mine for as long as we could. As the ship began to leave, my mom walked quite a distance along the pier, still holding her end of the streamer. We had a short stop in Japan and then went on to Hong Kong, where we disembarked. From there, we flew to Saigon and then on to Danang.

The first morning back when I started to slowly wake up, I heard the familiar *swish, swish, swish* of a broom as the servant swept the veranda. Even in homes where there are no verandas you hear early-morning sweeping. Grass is usually not allowed to grow around houses in Vietnam. Instead, the packed dirt around the area is kept clean and swept with a short, crescent-shaped broom made of wispy, long grass. When I opened my eyes, I surveyed the room through the gauze of the mosquito net. *I should get up. There is so much to do,* I thought to myself. Instead, I rested a few minutes longer, savoring the happy realization that after all our travels our family was finally back home in Vietnam. We would now be living in Danang.

It was April, 1967. Our two precious daughters were three and four years old.

Our first house

Learning Hrey

Village sorcerer

Buffalo sacrifice

Monotonous gongs

Toothless smiles

Distributing medicine

Learning to write

Literacy class

Eager students

Weang, his wife, and son

Old Rai

Bato, Vietnam

Christmas program

Hrey congregation

Tet Offensive

Vietnam has many special days of remembrance and celebration. Without question, though, the most celebrated of all is Tet, the lunar new year. In 1968, the North Vietnamese enemy forces decided to pull coordinated surprise attacks on many of the cities of South Vietnam. The Liberation Army hit hard with simultaneous attacks all the way from the demilitarized zone in the North to the delta in the South, which came to be known as the Tet Offensive. Every year, the Communists threatened to hit at Tet, but it never happened until 1968. The South Vietnamese and the American military were taken by surprise, and so were we.

On January 22 through March 31, 1968, a Wycliffe-sponsored literacy workshop was scheduled to be held in Kontum City, Kontum province in the highlands of Vietnam that shares borders with Laos and Cambodia. This workshop was going to be an excellent opportunity, as we needed further training from experts on improving our reading primers for Hrey children. A new-to-us method was going to be presented and explained with lectures and many valuable workshops, during which time we would be able to actually work on our primers and receive expert guidance.

We flew to the inland highland city of Kontum with our daughters, Ngiah, and his family. By then, Ngiah was our key man for the Highlander Education Program. We settled in at the Literacy Center in half of a double bungalow and began attending the extremely interesting sessions of the literacy workshop.

At midnight on January 29, we were fast asleep when suddenly the silence exploded. The Vietnamese soldiers and people were welcoming in their lunar new year with gunfire and firecrackers. We rolled over with a groan and went back to sleep. I think I developed my aversion to banging firecrackers about that time. I love the colorful sky displays on the Fourth of July but cannot appreciate the sound of firecrackers being set off just to make a loud banging noise, as it sounds to me too much like war.

At 2:15 a.m. on January 30, we were awakened again by the loud *rat-a-tat-tat* of machine gun fire. This definitely was not the sound of celebration. We yanked at the edge of our mosquito net that was tucked under the mattress and climbed out of bed. Still not quite awake, we hesitated, not knowing quite what was going on or what to do. Then the wail of the under-attack siren reached our ears, and we dashed into the other bedroom, snatched our two daughters from their beds, and proceeded into the kitchen. From there, we entered the other half of the double bungalow occupied by another missionary family, as the bunker was under their bedroom floor. Dick and Lillian Phillips had already gathered their three small children and opened the trap door. We all scrambled down the steps into the bunker.

We were surrounded by the loud, unmistakable noises of war. For a second, we sat stunned and just stared at each other. Suddenly, our almost five-year old-daughter Janice sat up straight and, lifting her arms with the palms of her little hands facing us, instructed in a loud voice, "Now, we're not going to think about the *emeny* but only about Jesus!"

War such as we had never heard raged above us. The enemy had attacked the military facility near the Workshop Center. The Tet Offensive 1968 had begun.

By 6:00 a.m., all was quiet, so we climbed out of the bunker and stumbled back to bed. At 7:00 a.m., a Huey helicopter landed in our front yard, waking us up. We were at the door in an instant to the knock of an American colonel who came by to see if we had any casualties. We were glad to report that no one had been hurt. He told us it had been a fierce attack, but they had held off the enemy all night.

He warned us to stay at the center and not go into town because there might be snipers around.

Since it was a holiday, we had no servants, so I washed some clothes. At noon, Oliver, another missionary couple, and I were almost finished hanging our wash on the line when bullets started literally zinging past our heads. We ran into the house, wondering what would happen next. Oliver and Dick Phillips were standing at a picture window when a bullet whizzed past them, went through an inside wall, and lodged in a door on the opposite side of the wall. We quickly got our children, grabbed some lunch from the kitchen, and scrambled back down into the bunker. At 3:00 p.m., we went upstairs to pack our language materials and some clothing in case we were evacuated.

We had previously invited another missionary couple over for the evening meal. I began to make preparations for their visit, and by 6:00 p.m., everything was ready. I even had time to make a lime meringue pie. (There were no lemons in Vietnam.) Suddenly, a helicopter landed in the front yard again, and several soldiers rushed out of the chopper with their M16s facing in all directions. Then, out jumped the chaplain, who told us we had to leave at once, as the area was under attack from the North Vietnamese army. This man had kindly convinced his commander to send a chopper for us. At first, the men on the chopper said they would take none of our tribal people. They were understandably wary of anyone but Westerners. When we insisted that we would not leave without them, they relented. We explained that we had brought these tribal people here to assist us and to leave them behind was unthinkable, as it would have meant their certain death. All the women and children were taken first, leaving all the men behind.

This was my first chopper ride. I recall the strange sensation of being literally lifted up. A Scripture verse came to my mind, which described just that: "The eternal God is your refuge, and underneath are the everlasting arms" (Deuteronomy 33:27). I was so concentrating on this truth that I forgot to worry about our situation. In minutes, though, we gently put down and quickly exited the chopper. The

Special Forces MACV compound was almost right next door to the Literacy Center.

Suddenly, American soldiers came toward us, handing us any food provisions they happened to get their hands on. Those American guys looked at us somewhat astonished, obviously surprised to have women and children suddenly appear in their midst at such a time. As for us, we were grateful and touched by their concern and thoughtful generosity. Some of the things they gave us were cans of tuna fish, chips, ripe olives, canned fruit, and peanut butter. We were then assigned and escorted to officer's quarters after being shown our assigned bunkers. Meanwhile, the chopper had gone back to the Literacy Center to get the men. No sooner had they returned, the wail of the siren sounded, and we all had to scurry to our bunkers.

Our bunker was not the usual underground kind but was above ground, made of sandbags with a doorway on one side. It was narrow, and we had to keep our knees bent as we sat with our backs against one side. We each held a daughter. There were several other missionaries in the bunker with us while other missionaries and tribal language helpers were in different bunkers. Altogether there were, as I remember, twenty-five adults and twenty-one children which included tribal helpers and their families.

For a solid twelve hours, the enemy was kept at bay. In addition to the sound of automatic weapons, mortars, and rockets exploding, we experienced a new nerve-racking sound. When an illumination flare was sent up to detect enemy positions, it made a terrifically loud, bone-chilling screech. We sat huddled together through the long night with these loud, unnerving sounds of war all around us.

We ate some of the rather interesting assortment of food that we had received upon our arrival. Eventually, the children went to sleep. We adults occasionally dozed off, and in between dozing and waking, we prayed a lot. The situation was tense and scary. We could only wait to see what would happen. God's Word came to my rescue and thankfully bolstered my faith.

> He who dwells in the secret place of the Most High shall abide under the shadow of the Almighty. I will say to

> the Lord, "He is my refuge and my fortress; My God, in Him I will trust.
>
> —Psalm 91: 1-2

Oliver remembered a different verse that revitalized his trust during this dangerous situation.

> A thousand may fall at your side, and ten thousand at your right hand, but it shall not come near you.
>
> —Psalm 91:7

An hour before dawn, things were still going strong. Even the chaplain and the cook were called out to the firing line. At one point, the chaplain checked on us and warned us to keep clear of the doorways. If the enemy overran the camp, the chaplain was hoping they would think the bunkers were empty. He was concerned that otherwise they would throw in grenades. The tenseness of the situation was palpable. Finally, full air support arrived and defeated the enemy just in time. We were told that in only a half hour more of fighting, the camp would have been overrun. As it was, the perimeter had been broken through a couple of times during the night.

The all-clear siren brought us out of our bunkers to stretch and enjoy the bright, warm sunshine. The enemy always withdrew during the daylight hours. We were treated to some delicious food in the dining hall, especially scrumptious steaks! It was surreal.

Oliver and some of the other missionary fellows wanted to return to the center to rescue some things. They were driven there in a military truck. A couple of the houses were in shambles. Our house was still intact. Fearing the possibility of booby traps, Oliver found a long pole and reached in and grabbed a suitcase near the door. He noticed that our dinner was still on the table, untouched. There was a pile of clothing thrown in the middle of the floor. The North Vietnamese had taken some clothing, our mosquito nets, and some cans of juice. Oliver grabbed some things and ran to the truck that brought them all back to the MACV compound at about 6:00 p.m. Just as they drove into the gate, the siren sounded once again, and we ran to our assigned bunkers for another night, January 31, of intense fighting. The enemy was attacking in human waves and was only stopped by the Americans aiming their big guns at them point blank. Still, more came.

We were awed by the sound of a C47 gunship that arrived, equipped with Gatling guns that could put a 50mm caliber bullet in every square foot the size of a football field in one minute. This gunship sounded like the roar of a huge animal. No wonder they were dubbed "Dragon Ships" and sometimes "Puff the Magic Dragon." We huddled together in the bunker in amazed astonishment!

When day dawned, we crawled out of our bunkers and once again enjoyed American military hospitality. Children are amazing. I noticed that some of them were playing a game of fighting, not between cowboys and Indians, but between GIs and VCs in the sand around the bunkers.

That afternoon, February 1, transportation was provided for us. We all climbed into the back of American military trucks, and we were driven through the streets of Kontum in an armed military convoy. Kontum had truly suffered the ravages of war. The marketplace and half of the city was destroyed. We arrived at the Kontum airport, which was in shambles. There was broken glass everywhere. All of us missionaries with our tribal assistants waited for an Air America plane to pick us up. Soon the C47 arrived, flown by an Air America pilot who had volunteered to rescue us. We seated ourselves in the sling seats along the sides of the plane and were flown to the coastal city of Nha Trang, where we were once again taken in an armed convoy, this time to the American army base, Camp McDermott.

There we learned that some missionaries in Banmethuot had been murdered, and others had been captured. It is hard to express the sorrow we felt for these dear friends who had come to Vietnam to serve, minister, and help people. We grieved with their loved ones. Eventually, we heard some of the cruel details of what had happened to each of the martyrs, which distressed us immensely!

One could ask the question, "Why?" Yet the question I pondered was whether or not Christ had fulfilled his promise to be with them, as they had been killed. Was He with them? Of course He was! But

He was not with them as we think He should have been to keep them alive as He had done for us and for many others. And He could have done just that, of course. It is impossible for us to know what these innocent people experienced in their last painful moments of life on Earth, but this I believe with unshakeable certainty: Christ was with them in their last moments of unbelievable shock and pain, and then they were with Him as He welcomed them home.

We read in Scripture of so many others who died as martyrs. Think of John the Baptist who was imprisoned for daring to speak out against King Herod for his unlawful marriage. Herod's wife got her revenge when drunken Herod, being pleased with the dance performance of the daughter of his wife, promised her anything her heart desired. The girl's mother advised her to request the head of John the Baptist. Imagine John, there in a prison dungeon. Suddenly, guards arrive, and he is taken out and beheaded. There is no record of the last moments of this illustrious, faithful servant of God. We read in Scripture of many others of God's faithful servants who were also summarily executed.

There is one very interesting account in the Bible, though, that gives one pause. Stephen, a deacon of the early church, had the luxury of preaching a clear, impassioned sermon of devastating truth before he died the painful death of stoning by an angry mob. This time, we get a glimpse of someone's last moments on Earth before being so unjustly killed:

> When they heard these things, they were cut to the heart, and they gnashed at him with their teeth. But he, being full of the Holy Spirit, gazed into heaven and saw the glory of God, and Jesus standing at the right hand of God, and said, "Look! I see the heavens opened and the Son of Man standing at the right hand of God!" Then they cried out with a loud voice, stopped their ears, and ran at him with one accord; and they cast him out of the city and stoned him. And the witnesses laid down their clothes at the feet of a young man named Saul. And they stoned Stephen as he was calling on God and saying,

> "Lord Jesus, receive my spirit." Then he knelt down and cried out with a loud voice, "Lord, do not charge them with this sin." And when he had said this, he fell asleep.
>
> —Acts 7:54-60

Such I believe was the caliber of the six selfless missionaries who were mercilessly and senselessly murdered during the 1968 Tet offensive. There were two other missionaries captured at that time who died after suffering months of sickness, mistreatment, malnutrition, and hardship on lonely, wet jungle trails. There were three missionaries from as far back as 1962 who had been captured and never seen or heard from again. In 1963, two missionaries and a child were murdered on the road. Then in 1966, one of our WEC missionaries, John Haywood, was gunned down while on an errand of mercy. These martyrs were the Stephens of that war who saw the heavens open and Christ standing to welcome them home.

Camp McDermott in Nha Trang was a whole new experience for us. We lived there for a month in Quonset huts lined with double bunks. We put two double bunks together with our girls on the bottom and us on the top, as this would give them more protection, and we could just jump down and grab them if necessary. And it was necessary a couple of times. The siren would wail, and we had to quickly walk a short distance to large underground bunkers. However, there was a problem. Since we were obviously not American soldiers, we could be mistaken for the enemy and get shot. We were warned therefore to not leave our huts when the siren sounded until an armed guard came to escort us to the bunkers. For the most part, we felt confident that we were pretty safe at Camp McDermott. Still, my legs felt weak and shaky whenever we had to hurry to a bunker, which was not the case in the much more dangerous situation we had been in at Kontum.

Later, we learned of how we had again been saved from certain death at Kontum. There were six 122mm rockets discovered that were set to hit the MACV compound where we were huddled together in bunkers. However, for some unknown reason, the timing device failed to function.

We enjoyed the food at Camp McDermott. Perhaps soldiers sometimes complain, but I was pleased with the food we were served, with much of it flown in from the United States. Our children especially loved the seemingly endless supply of chocolate milk.

Some stateside mission leaders of various groups were understandably concerned about everyone's safety. We were concerned about our safety, too. However, not one missionary that we knew of was willing to leave Vietnam in 1968, as we sensed it was not time for that yet. So we stayed.

Oliver often prayed, "Dear Lord, thwart the enemy's plans." We had to smile when Jeanette took up the refrain and prayed fervently, "Dear Lord, swat the enemy!" That is exactly what had happened in 1968. The enemy had been repelled, the South Vietnamese Republic held, and life continued pretty much as before. But by now, though, we were reluctantly beginning to suspect the day would eventually come when we would have to leave. How we dreaded the thought of South Vietnam falling to atheistic communism. Meanwhile, we were able to carry on with our work.

A two-story house was rented in the city of Nha Trang. We all went there, where we were assigned bedrooms, and the Wycliffe Literacy Workshop continued to function beautifully. Our coworker, Pam Brady, even joined us from Danang and she kindly helped keep the children busy while we adults attended workshops and worked on our primers. We received tremendous, valuable training and returned to Danang in July of 1968, where life was back to normal, Vietnam style. We were all safe and busy at home once again in Danang.

Danang

We now made our home in Danang because we were not allowed to return to Bato. It was really different living in the city rather than in our lovely valley surrounded by jungle-clad mountains near the Hrey villages. But it was so nice to have electricity, though, which we did not have at Bato.

We were able to rent a large house. At first, we thought it was too expensive for us, but then suddenly the piaster was devaluated, and the monthly payments became one-third of what they would have been. This three-bedroom house was perfect for us and for our coworker, Pam. The living room was large. We divided a section of it off with tall bookcases, which became Oliver's office. There was also a dining room and an excellent Western indoor kitchen with a pantry! Adjoining the main house in the back was a long room, which made perfect bedroom/office for Ngiah and his family. There was even a small bedroom with a private entrance for our live-in Vietnamese servant. One whole side of the house was paved, and there was a small, narrow garden area. A cement wall surrounded the whole property, with a wide, large metal gate at the entrance. We really liked it.

We invited Ngiah, his wife, E-Ne, and their daughter to come to Danang to be part of our team. Since their daughter was their firstborn and was named E-Tsee, among the Hrey, Ngiah was now known as *Baq E-Tsee* (Father of E-Tsee), and E-Ne would be called *Miq E-Tsee* (Mother of E-Tsee). Since they were now living in Danang, which was a totally Vietnamese environment with no other Hrey people, they preferred to be called by their given names, Ngiah and E-Ne.

We got right to work preparing materials for the Highlander Education Program (HEP), which was the South Vietnamese government program being developed for tribal children to learn to read and write, first their own language, and then move on to learning to read and study in the Vietnamese language. Other missionaries were also involved in this program.

This became a huge undertaking. Along with the three reading primers we were developing to teach tribal children to read and write their own language, we were also responsible for developing detailed teacher's manuals for each primer. We also had to translate Vietnamese into Hrey elementary school books on science, math, ethics, health, and prepare guides in Hrey for teaching those subjects.

After having completed these Hrey materials, it was our responsibility to hold teacher-training classes for Hrey men and women. First, we had to recruit potential teachers, teach them to read and write their own language, and then to follow the teacher's guides in order to teach the children correctly. Each book had to include a summary of its content in the Vietnamese language.

A fun project was having Ngiah tell us some of their folklore stories to be made into a book, which was also a requirement of and printed by the HEP program. Of course, these stories had never been written down and had only been shared orally. It certainly was an exercise of viewing some things quite differently.

Ngiah and I had joyful time working together on the translation of forty hymns into Hrey, which was not part of the government program. It was quite a project, as it was the first Hrey hymnal done with music notes.

Sometimes we visited wounded Vietnamese and Hrey soldiers at the Danang US Special Forces hospital. We purchased towels, washcloths, combs, toothpaste, toothbrushes, soap, and candy. Then we set them out by categories around our living room, and our girls would fill the plastic bags with these items. When they were finished, we packed them up and went to the hospital. Our girls loved handing these Care Kits to the wounded patients. Each patient smiled his appreciation as they gratefully received their package from our little blondes. We were able to give out a total of 1,250 kits over a period of time.

We designed a U-shaped desk for homeschooling, which was made by a skilful Vietnamese carpenter. I sat in the middle of the U with a daughter on each side. Their bedroom was large, so half of it became our schoolroom. Janice started kindergarten, and Jeanette began with some preschool projects using materials I prepared as needed. We were enrolled in the Calvert School system, which was used by families of diplomats around the world. Receiving all the supplies to start our school was exciting.

One day after some weeks, Jeanette said, "Mommy, I think Janice is in real school, and I'm in pretend school." Oh dear, she must have realized I was supplying coloring and other preschool projects for her, which obviously were not as impressive as the Calvert materials. Things were much better when Janice was in first grade and Jeanette in kindergarten. Then everyone was content and busy in "real school."

Our school each day finished at lunchtime. After lunch, everyone had a siesta for about an hour when we either napped or read. I have often wondered, of all the things we borrowed from other countries, why didn't we borrow the siesta? And in hot, humid Vietnam, with no air-conditioning, siesta was a special blessing and more of a necessity than a luxury. By this time, though, we did have ceiling fans, for which we were thankful.

After siesta, the girls were always busy and creative in thinking up ways to entertain themselves while we worked on various educational materials for the Hrey. Once, they made the front porch glider into a beauty shop. The two customers were Ngiah's two daughters, who had their hair fixed several different ways. Sometimes the four of them would spend time together coloring in books that Grandma sent from America, playing on their slide and swings, or with a toy doctor kit. Our girls would tend the "wounded" E-Tsee and E-Tsai in the shade of our bamboo trees. We noted how seriously Jeanette took her pretend medical profession as she took blood pressure, listened to heartbeats, gave shots, and wrapped pretend wounds.

We decided to make Friday nights Family Night, since in Danang, there were no places we could go for entertainment. There were no skating rinks, no swimming pools, or malls. Also there was a curfew. So we made our own family fun, mostly by playing table games together. We saw how much our children looked forward to that time each week when we put everything else aside to spend a time of fun and relaxation with each other.

Many interesting items began appearing in the markets of Vietnam. One of these items, which seemed to be in abundant supply, was the American military C-ration chocolate candy bar, which became our Friday night special treat. Vietnam had been a colony of France for one hundred years, so we noticed a lot of French influence in our early years in Vietnam. It was interesting to witness a shift toward American preferences, and people became extremely eager to learn English.

We ordered a metal-framed, plastic collapsible pool from Sears via APO. All the children loved cooling off in the water. It even had a slide on one end

We were given a couple of darling white rabbits. Soon, there were several darling baby bunnies that the children enjoyed dressing in doll clothes. Sadly, one of them became sick and had to be put down. While Oliver was dealing with that, I took the girls into our bedroom. We had a three-person cushioned, wooden bench that our carpenter had made for such occasions and where Oliver had devotions with them every evening. Now they were upset about the sick rabbit. We huddled together, feeling sad. I explained to them that animals are wonderful creatures created to give us pleasure. However, animals are not people. People are created in the image of God and are eternal, whereas this is not true of animals. The death of a person is extremely sad. The death of an animal, though sad, is not nor should it be the same. Since they were mostly sad because the poor thing had been sick, I explained the importance of putting it out of its misery. They agreed. We sat and talked for a bit, and then they went back outside to play.

There were important projects to keep Janice and Jeanette busy, like carefully creating birthday and Christmas cards to mail to loved ones, especially to Granddad and Grandma in America. We had a hand-crank ice cream maker, and my mom sometimes sent us some wonderful ice cream mixes. One evening, while we were enjoying this special treat, Janice declared, "I am going to make a thank-you card for Grandma for this ice cream. She is so wonderful!" Yes, the ice cream was a fantastic treat. The card was carefully designed with a message of gratitude and sent off.

We had many small parties to celebrate birthdays for our family, Pam, Ngiah, and his family. I always tried to have some sort of special theme. Once, we invited some children from another mission organization to help celebrate. We put two chairs in a row, side by side, in the living room to represent the passenger area of an airplane. Oliver was the captain and, through a small loudspeaker, made appropriate pilot takeoff and landing comments from his office on the other side of the bookcases. Birthday refreshments were served to the passengers on trays.

We enjoyed a lot of good fellowship with Christian American Air Force officers and enlisted men in our home, who often did special projects for us. They had to get permission to leave the base and come to our house. They built kitchen in the back for Ngiah's family and our servant girl to use, as well as a garage at the side of the house. They were the ones who erected the swings and slide for our children.

One time, they brought us an instant water heater that they installed in our shower, which was a real luxury in the cooler months. It is surprising how cold we felt during the cooler season because of the humidity, and we welcomed those warm showers. After each project was completed, we served the men a meal. One time, they were having a problem with something they were working on, and it got too late for them to eat dinner. They had to rush back to the base before curfew but not before they scooped up some cookies and filled their pockets. The cookies were probably my "famous" chocolate chip-less cookies.

One day in Danang when we came out of a store, and were trying to get back to our Land Rover parked in front, we were surrounded by scores of friendly, curious, Vietnamese children and adults wanting to look at our daughters and especially all wanting to feel their blonde hair and squeeze their cheeks! They kept blocking our way. The girls were becoming agitated. Just then a couple of GIs came to our rescue e backed off. We got back into the Land Rover and invited iends (our heroes) to our home for a meal sometime. They re than once. We thoroughly enjoyed having fellowship with them.

We had three languages going at all times at our home. We usually spoke English between ourselves and Pam, Hrey with Ngiah and his wife, and Vietnamese with Phuong, our live-in servant. Sometimes I got mixed up. A couple of times when I was going through the schedule and menu for the day with Phuong, she looked at me strangely. Finally, I realized that I was speaking in Hrey and had to do the whole thing all over again in Vietnamese. There were several such amusing incidents.

Phuong came to us when she was eighteen years old. In those days, the going rate for someone like her was about twelve dollars a month. She was a good worker and Christian friend to us, who had some serious concerns and asked us many questions. Often we would stop and share Scripture truths with her as we went about our daily activities. These were not Bible study classes; we just answered her questions whenever she asked throughout the day.

One time, Phuong was confronted with a serious problem concerning her mother who was a widow and who professed to be a Christian. However, she insisted on visiting her husband's grave and performing certain pagan rituals. Phuong was troubled by this and shared her concern with us. She refused to have any part in the offerings to her dead father except to occasionally weed his gravesite, and this made her mother angry.

Back in her village, Phuong's mother began having some horrible problems. She would scream uncontrollably, literally climb the inside walls of her house, and tried to throw herself down a well. Finally, the village people chained her to her bed. They then called Phuong home to her village.

Phuong later reported that she was unable to do anything with her mother. Eventually, she did settle down enough so she could be unchained. Phuong felt absolutely helpless, so she decided to fast and pray. Her mother did not want her to read her Bible. Phuong would go outside and sit secretly in a corner of the pigpen, but her mother would find her and throw stones at her. Phuong fasted for six days. On the seventh day, she decided to lay hands on her mother and pray for her, which she did. Her mother was completely transformed and came back into her right mind.

Later, Phuong's mother told her that wild, demon-like animals had surrounded her, threatening to bite her if she did not scream, and that the only time they did not come into the house was when Phuong was there. Then her mom said she saw a man with a shepherd's crook leaning on the side of the doorway with one leg blocking the way. We rejoiced with her in this wonderful victory.

Phuong returned to continue working for us. She made delicious Vietnamese meals. Preparing a proper, nutritious Vietnamese meal involves a lot of time because of the necessary chopping of vegetables and the special preparation of meats with all the delicious sauces. We had a couple of those meals a week. Other times, we enjoyed simple fried rice or Vietnamese rice-noodle beef or chicken soup (*Pho*) with rice-flour crisp bread and peanut butter, which we made by grinding peanuts in our hand-operated meat grinder twenty-five times to get the desired creamy consistency.

The most wonderful, fresh French bread is available in Vietnam. The French taught the bakers how to make it during the colonial years when Vietnam was part of French Indochina. The only bread available in Vietnam, it was sold from the back of bicycles every morning, freshly baked and still warm.

We also enjoyed Western meals, which usually meant larger pieces of meat that required the use of our pressure cooker to ensure tenderness. Someone told me once that the reason Asians cut their vegetables and meat into small pieces is because in China the cook was also the one who had to gather the wood for cooking. Smaller pieces of food take less cooking time and therefore less wood.

In Danang, there was always a large assortment of fresh vegetables to choose from at the open market along with a variety of delicious tropical fruit. We almost always preferred steamed rice even with Western meals, but sometimes we had potatoes, which were also available. There were no fresh dairy products in Vietnam, so we bought powdered milk imported from Holland at a store owned by a Chinese family. We also bought butter in cans there.

I developed a recipe for pizza when I discovered we could buy cheese in cans and Spam at that same store. Then all that was involved was to cook some local tomatoes, add pizza spices sent by my mom, fry some Spam, pile it all on French bread, top it with cheese, and bake until the cheese melted. We thought it was delicious, especially since we did not have authentic pizza around for comparison. It is rather amusing that our Vietnamese cook could make Italian pizza by topping French bread with cooked local tomatoes, spices from America, Spam from Korea, and cheese from Holland purchased at a Chinese store.

Phuong became quite an expert cook. I also taught her how to bake cookies, pies, and cakes, all from scratch, using recipes from my trusty Betty Crocker Cook Book. I translated them into Vietnamese, and soon we had a handy American recipe book written in Vietnamese for her to use. So Phuong did the cooking and baking. The downside of this is when we went on furlough and I offered to make dinner one night while we were staying with my mom, Janice exclaimed, "Mommy, can you cook?"

Inviting our American military friends over for holidays like Thanksgiving and Christmas was always a pleasure in Danang. They were so appreciative, and we had some good times together. Whenever I was with them at the US Air Force base, the beach, or Country Church, I prayed that when they saw our family they would be reminded to be faithful to their wives back home.

I especially remember one time when we invited two officer friends to have dinner with us one evening. Phuong and I went to great lengths to cook some really special dishes. We wanted to treat them to some of the delicious foods of Vietnam, which we knew Americans would enjoy. We set the table, and all was ready. We waited and waited, but they did not arrive. Oliver drove out to the air base and finally found them as they were coming out of their dining room. They had already eaten! However, they were so sorry that they had forgotten, and they still wanted to come. They bravely partook of some of our dinner. The evening turned out to be enjoyable, in spite of the mishap. These guys were real gentlemen!

A couple of the Christian officers discovered how much Oliver loved ice cream, which was not available in Vietnam. Often after the Sunday night service on the base, they invited our family over to the Danang Officer's Open Mess, the DOOM Club. There they tried to get as many tribal adventure stories out of Oliver as possible as they plied him with bowls of ice cream, from America. Somehow he never objected!

Once when we were having fellowship with some servicemen, one of them, Hoyt Richardson, drew a caricature of the chaplain on a paper napkin. We needed an artist to illustrate the scores of pictures in the three primers we were developing. Hoyt became an invaluable asset to our ministry as he faithfully made many excellent ink sketches for our reading primer. One time, we wondered how to depict a Hrey person who was laughing. Ngiah demonstrated by throwing one hand in the air while repeatedly slapping his thigh with the other hand. Hoyt captured that pose in one of his many drawings.

There was one serviceman, Earl Kilpatrick, who was always ready and eager to do things for us. Suddenly, he disappeared. No more offers of help, no more help, no more Earl. Then one Sunday morning after we returned home from the local Vietnamese morning church service, Earl appeared at our house in the base commander's car. We were somewhat shocked but also impressed. He informed us that the base commander wanted to see us, and that we should all come immediately with him to the base. We were puzzled but obediently piled into the car and were driven to the base, where we were escorted into the commander's office and invited to sit down. After a few minutes, the commander came in and sat behind his desk. He wanted to know about our work as missionaries. Then he asked, "Did you see that black Scout jeep in the parking lot?" I don't think we had. Suddenly, he pushed keys across the desk and said to Oliver, "It's yours!" Oliver was speechless! We learned later that Earl and his buddies—one of them was Rex Cogar—had scrounged for parts from five different vehicles and had been busy for weeks putting together this one. It was even freshly painted. That was a wonderful provision for us because up to then we had gone to the base on a Honda 50 motorcycle. I would sit behind Oliver while Janice sat behind me and Jeanette sat in front of him, which was not unusual because we saw as many as six on one motorcycle. The timing of this new-to-us vehicle was special because that same day happened to be Oliver's thirty-fourth birthday—and the men did not know it! You can imagine how happy that made Earl, his buddies, and even the base commander. As for us, we felt honored and privileged to have such friends. There are probably hundreds of such stories of American servicemen helping and blessing us and many other missionaries in Vietnam. We were incredibly saddened to see how Vietnam veterans were mistreated after returning home, especially since we and other missionaries witnessed so much goodness and received such help from them. We know that many of them were responsible for helping orphanages, churches, missionary projects, and the list goes on.

Our pastor, Reverend Paul Fryhling, visited us before this and had ridden in the Scout, which he dubbed, "The Black Mariah." We took him to lots of places in Danang while he was visiting. One time, he wanted to take a picture of something but needed to climb onto something to get a good shot. He returned to the car, looking somewhat disheveled. He had fallen but was not hurt. His comment was, "God sure takes good care of dumb people." Ever since then that has been one of our family's handy quotes when necessary!

When he returned to America, Pastor Fryhling encouraged First Covenant Church in Minneapolis, to buy us a vehicle, a blue station wagon that was shipped to us from Japan. It was even air-conditioned! What a luxury!

One time, I decided to introduce Ngiah to the fun of April Fool's Day. There was always concern about thievery in Vietnam, and our Honda motorcycle was stolen one night. The station wagon was kept in the garage built by our Air Force friends, and the Scout was parked beside our bedroom window. I coached Ngiah to get up extra early on April first and start shouting that the tires were gone. Early that morning, he cried out in Hrey, "Oh, missionary, oh, missionary! Come quickly! The tires are gone! Someone stole the tires!" Oliver was on his feet before he was fully awake and shot through the dining room, through the kitchen, and out the door. It took him some minutes before he realized it was not true as he walked around the vehicle. By this time, Ngiah was at the other corner of the house laughing his head off while I shouted "April Fool!" Oliver, realizing he had been tricked, squinted in mock anger and shook his finger at Ngiah with a gesture of "You will be sorry for this!" We all had a big laugh.

And payback time did come. Oliver had rigged up an alarm-bell system with a trip wire in the front of the house to scare off any would-be thieves. He also had installed an alarm bell on the side door of the garage, which would be set off if the door was opened. The switch for it was in our bedroom, which we turned off during the day. Sometimes when Ngiah opened the garage door, Oliver would turn on the switch, and Ngiah would just about jump out of his skin, to the sound of Oliver's hearty laugh. A couple of times, Oliver turned it on just as Pam was getting her bike out of the garage to go on her evangelism and Bible-sharing visitation rounds to Vietnamese. He loved to hear her, "Ohhhhh, Oliva (British for Oliver)!" Then, grinning, she would go on her way.

We were glad to give our Scout to a Vietnamese orphanage in An Khe, which was a contact through the aunt of one of the Vietnamese pastors. They were grateful with this provision, and it gave some years of useful service to them. We heard later that it had been somewhat damaged in an accident but had been repaired and was back in use again by the orphanage people.

One day, Ngiah left a note on the office desk. Oh-oh! We knew that there was some problem he wanted to bring to our attention, as this was his way of doing so. He would write us a note, his reasoning indicating there was a serious problem. I would get a knot in my stomach, wondering what could be wrong. This time, the note was to me because he felt I had to do something about his wife. The translation from Hrey read, "Today, I want to request that you, Miq E-Jan, remove completely all of the stones in the ears of Miq E-Tsee! And also you must remove them from her eyes!" We learned she had locked him out of their bedroom. This made me smile, although it was quite shocking. Poor Ngiah! He obviously wanted me to bring the life-changing gospel to E-Ne so she would be a better wife and thereby make life easier for him.

We never learned what the problem was between Ngiah and his wife, but she must have been really upset about something. Actually, she always seemed somewhat sad. She was probably barely twenty years old, had been brought from her tribal village to live in a strange city, and all this was no doubt new and difficult for her. At any rate, I was firmly commissioned to teach her Bible truths and thereby remove stones from her ears and eyes! Now, this was an interesting and challenging assignment, to say the least. I decided I would do this during siesta.

Once a week, E-Ne and I had our class together. I had learned at Bato that tribal women are usually not interested in book learning. Back at Bato, when I taught some of them one clearly stated fact and then immediately asked them a question about what I had just taught, they usually answered, "*Uh nee*" (don't know). I soon realized that either they did not know, did not want to know, or did not want to admit that they knew, as being women they were not supposed to know. I never learned what the real problem was.

With E-Ne, I used Sunday school pictures sent from home and thoroughly enjoyed telling her Bible stories. I assigned her a short verse to memorize each week, which she faithfully accomplished. Usually, I had to translate the verse that went with the Bible story and then check with Oliver and Ngiah for accuracy, as there was as yet only the gospel of Mark.

When I started this class, E-Ne brought their youngest child with her, wrapped in her ever-present and useful tribal baby-carrying cloth. Naturally, if the child was asleep, all went well. If the baby woke up, and sometimes I suspected that E-Ne purposely woke the child, that would pretty much be the end of the class. I finally told Ngiah that if he wanted me to teach his wife, he would have to keep the baby with him during siesta. This he was willing to do, so teaching became much easier. My routine was that I always opened with prayer, and she would recite her assigned Bible verse. Then I would share a Bible story with pictures and ask questions to check her comprehension. We sang a song together from the hymns Ngiah and I had translated into Hrey, and then I closed the class with prayer. It was so gratifying to me when she was able to answer questions and began to show some interest. She was also learning to read, which gave her a sense of achievement. She expressed her happiness with her wide, toothless smile and shining eyes.

Sometimes, though, she really upset me. We had invested in some potted flowering plants, which brought welcome color around the outside of the house. E-Ne would allow her children to break off the leaves or even the flowers of my precious plants. I never did mention it to her, but I sure did not like it. One day, I was convicted about my bad attitude when it finally dawned on me that I had not come to Vietnam to nurture plants. I feel ashamed to admit that I even had to write in my journal an important reminder, "Do not get upset about plants!"

The weeks passed. I wrote to our faithful prayer partners in America and Australia to pray for E-Ne, which they faithfully did. We were always keenly aware that our praying partners were a vital part of our ministry. Then one day, E-Ne and I had our usual class time together, which didn't seem different or special at the time but the outcome was splendid. We continued with our usual routine, which always ended with my closing in prayer. When I lifted my head, I realized that shy, quiet E-Ne still had her head bowed and was murmuring words. This had never happened before. She was speaking so softly that I could not quite hear what she was saying. Realizing that she was not talking to me, I waited until she finished. She looked up at me, smiled shyly, picked up her papers, and left. I mentioned to Oliver that I thought she had invited Christ into her heart, but that I was not really sure.

A couple of mornings later, I heard someone outside our bedroom window. I peeked out and discovered much to my utter amazement that it was E-Ne, watering my plants. I was surprised and impressed. Later, Ngiah informed us that his wife wanted to help me in some way. I arranged it so that on the days that Phuong went to market, E-Ne could wash the breakfast dishes, which would be a big help. First, I had to teach her to wash her hands carefully with soap and water. Then, I put an apron on her. We got her a stool so she could properly reach the sink. I have a picture of that dear, smiling woman working away, joyfully doing our dishes, often with a baby tied on her back. My joy was complete when Ngiah quietly informed us one day that his wife did not scold him anymore. He was impressed, and I was overjoyed. My faith in the power of the gospel of Jesus Christ to really change people was certainly strengthened. This was a simple tribal woman who had been profoundly transformed.

As a young missionary, I had often heard it said of someone who was supposed to have become a Christian, "He prayed," or "She prayed." Yes, E-Ne had prayed, but the reality was that she began to demonstrate proof of having been changed. That is the fruit of the power of the glorious gospel.

There was another change that had taken place in E-Ne. Once in Quang Ngai while a Hrey friend was visiting her, there were some incoming enemy rockets. They had to run with their children to the safety of the bunker that was part of the house we were temporarily renting at the time. E-Ne informed me with a big smile that she had said to her friend, "I am not afraid. The Lord Jesus is with us!" I marveled at her grasp of that wonderfully profound truth, "Lo I am with you . . ." (Matthew 28:20b)

We had been working harmoniously with Ngiah on all our Hrey material preparation for many months. But then he started to change. In any kind of translation, you must have full mutual trust and cooperation. When Ngiah began grunting his approval at any of our suggestions and was obviously not paying attention, we became frustrated. Suddenly, he seemed to think we were holding out on him and he began acting strangely and checking carefully any provisions we had in the house. He wanted more wages although he was receiving the correct salary. One day, I was in the yard searching for a misplaced toy and asked Ngiah if he had seen it. He became extremely upset and accused me of accusing him of stealing. Nothing was further from my mind. Things became very tense.

Pam and I had had it! We thought Ngiah should be sent back to his village. But Oliver was much calmer and reminded us that there was no one else who could give us this kind of help. Besides, Ngiah had been our friend for some years by this time, and anyone else we might suggest to help us had been drafted into the army. Ngiah was available to us and not in the army because he was blind in one eye.

The situation was quite serious, so we three decided to fast and pray. Ngiah knew nothing of this. The next morning, he came in to work on translation with Oliver. I got a knot in my stomach when he passed the bedroom where I was teaching our girls because I knew Oliver was going to confront him about his attitude. When Ngiah got to the translation desk in the office, Oliver pushed the books aside and told Ngiah that they had to talk. Before he could say another word, Ngiah said, "I know, missionary. I am the problem." The ground of his heart had been prepared, and he was able to receive the help he needed from Scripture to experience forgiveness and deliverance from covetousness and gain contentment and peace by surrendering his self-will to Christ. The change in him was truly dramatic. Now we moved ahead with precious oneness that could only have been accomplished by the work of God the Holy Spirit. So we all rejoiced together. We never had another problem with this dear brother in Christ.

I should mention one other incident with Ngiah. It was a Sunday morning after church. Ngiah asked Oliver to come to their room for a minute, which seemed rather strange. When Oliver got there, Ngiah was standing by his desk, which was right inside the door, staring down at it. Finally, Oliver noticed a receipt from the Vietnamese church for a gift from Ngiah laying on the desk. Neither of them mentioned it. They made small talk, and Oliver went back into the house. The next day, Ngiah came in, upset. He had just been reading these verses from his Vietnamese Bible:

> But when you do a charitable deed, do not let your left hand know what your right hand is doing, that your charitable deed may be in secret and your Father who sees in secret will Himself reward you openly.
>
> —Matthew 6:3-4

Ngiah confessed that his reason for asking Oliver to his room the day before was so that Oliver would see that he had given money to the church. He felt bad about it yet had learned a wonderful biblical lesson. It was refreshing to witness how seriously he took the Bible's teachings.

Another time, we learned that each time Phuong killed a chicken for our dinner, Ngiah was always there to collect the blood. Oliver called his attention to Acts 15:20, which, among other things, speaks of abstaining from eating blood. Oliver then decided to do some research on the verse and wrote to a couple of scholars in America. When he started to share some of their explanations, Ngiah looked at him in wonderment and said, "It's okay, missionary. I stopped eating blood as soon as I read about it being forbidden in the Bible." Case closed.

Some months later, Oliver was going to visit a Hrey village, and Ngiah was going with him. They would have to be flown in by helicopter, as the roads were extremely dangerous because of anti-government activity. Ngiah had an uncle in this particular village, and he thought long and hard as to what might be the appropriate gift to bring to him. As he was discussing this with us, he commented, "I live like a king compared to him." Ngiah finally decided on the most precious thing he could think of, something extremely essential that would be greatly appreciated: a bag of salt.

Maybe Danang was considered safer than Bato, but sometimes it didn't seem like it. The Communists would often shoot off two or three rockets in quick succession in the night. They had to then scurry away from where they launched them, as our planes would be up in the air in seconds and could pinpoint them quickly. We had taught our daughters to carefully roll off the bed when the siren sounded and get under it in case something hit close by and things would fly. We knew that if we could hear the whine of a rocket, we were probably safe, as it was passing over us. One time, a rocket hit a house not too far from ours and wiped out a whole family. Another night, there was a huge explosion and a gigantic fire at the American air base because the ammunition dump had been hit. It lit up the sky. Oliver, much to my dismay, got up on the roof to take pictures. Although it was a magnificent display, it was also quite awful. The sound of explosions lasted all day.

Along with other ministry projects, we were working on some intense further checking of the gospel of Mark that we hoped to get printed. Sometimes translation work is especially complicated when clarifying who or what is happening to whom in certain verses. We grappled with one verse in particular:

> And whoever causes one of these little ones who believe in Me to stumble, it would be better for him if a millstone were hung around his neck, and he were thrown into the sea.
>
> —Mark 9:42

During the checking by a Wycliffe translation consultant of this particular verse, we discovered that the translation in Hrey said that it would be better for the child if a millstone were hung around the child's neck and the child would be thrown into the sea before someone caused the child to stumble! This was unacceptable. There was a lot of interesting discussion with Ngiah to get it corrected. When we finished it to our satisfaction, though, we still had the slight suspicion that Ngiah felt the first translation was more logical. We learned again the importance of checking, double-checking, and then checking again for accuracy in translating Scripture, which is extremely time-consuming work. It was important to have Ngiah re-phrase verses after we had already spent a lot of time carefully explaining them to him to make sure that the verse was accurate. We finally did get Mark printed.

One big interruption was when Ngiah became ill. He kept complaining that his stomach really hurt. When the pain became intense, we took Ngiah to the West German hospital ship in the Danang Han River. It turned out he had a ruptured appendix and had to be operated on immediately. After a few days, we went to check on Ngiah, and as we drove up to where the boat was docked, we saw him walking down the gangplank. He had just that moment been discharged.

A special event was when Oliver's brother Ted, his wife, Muriel, and one of their daughters, Cheryl, came to visit us in Danang. One evening, we invited an American Air Force friend, Major Lee, who was an F4 Phantom Jet pilot, to join us for dinner. After the meal, we sat in our living room, and Major Lee began sharing some of his hair-raising experiences as an Air Force pilot, which had us on the edges of our seats. Suddenly, someone near our house launched a military illumination flare just for the fun of it. They make a horrible screeching, nerve-jarring sound when they are sent up. We had instructed our relatives that if a warning siren sounded, they must get down on the floor because rockets had started hitting the city. The screeching sound of that flare shook us all, but we knew what it was and just sat there. Ted was halfway down to the floor but stood up again when he saw that we were not alarmed. We realized again that those of us who lived there had developed an awareness of war sounds that alerted us of when to get down. We all empathized with Ted's instant reaction and complimented him for his swiftness and agility. If the truth were to be told, though, we thought it looked quite amusing, and I am sure we were smiling quite broadly.

Our family all loved going to the US Air force base in Danang on Sunday evenings to attend the services that were held in a chapel on the base dubbed "Country Church."

Country Church

When we opened the door of the white clapboard chapel, a puff of cool air rushed out to greet us, instantly releasing us from the clutches of the hot, humid summer evening. We quickly made our way to our favorite pew, our two girls and I. Daddy would join us in a minute. He was busy chatting with some of the guys in the back of the church. He had brought books to sell after the service that might help them in their Christian walk. Soon, organ music filled the room and drowned out the sound of the air conditioners. The organist was also an Air Force dentist, and we were privileged to have him as our dentist, free of charge. Another wonderful provision.

The all-male choir dressed in military fatigues would file to the front of the chapel. With unconcealed enthusiasm they sang, "We have heard the joyful sound, Jesus saves, Jesus saves." This was the theme song of the evening service known as Country Church. The congregation of US Air force officers and enlisted men plus us missionaries listened with obvious agreement to this musical declaration. It was Sunday night at the US Air Force base just outside Danang, Vietnam.

One of the highlights of the service was the "Where It's At" session. As the testimony time began, one GI after another would jump to his feet to give praise to God. The men praised Him for help in the current happenings in their lives, how He was blessing their families back home, and for giving them courage and strength. This was especially meaningful to us, knowing the onslaught of temptations these young men had to face and the loneliness they experienced. We enjoyed hearing the men tell about the reality of Christ in their lives.

Some Sundays, we heard the startling, "I praise God for Vietnam!" Praise God for Vietnam? Who could do that? Wasn't the whole thing negative? Evidently, not to some of them. We listened eagerly as they explained that because of coming to Vietnam they had submitted to the lordship of Jesus Christ. They testified of His forgiveness, and that He was now their Lord and Savior. A variation of this testimony was, "I came back to Christ in Vietnam." There were many other words of rejoicing as men told of opportunities to share their living faith with their buddies. Other fellows spoke of Christ and His power to give them victory over or the grace to flee from powerful temptations.

The chaplain then treated us to a sermon. We were especially pleased that he was a sincere man of God who stayed true to God's Word. His sermons were biblical, good, sincere, and timely, which blessed and challenged us all in our walk with Christ. Sometimes men would search him out after the service and find forgiveness and salvation in Christ.

All too soon, the service was over. As we sang the closing hymn, it was a blessing to hear the sound of sincere mostly young men praising God.

The meeting always ended with a Fellowship Circle. How the men loved this tradition. We all joined hands and formed a large circle around the altar, to the back, and around the whole sanctuary. With reverence and contemplation, we sang together, "God be with you 'till we meet again." Standing together with those men in that war-torn land was poignant, as we realized that some we would not meet again here. And some did not return from their missions. Yet the sorrow was somewhat tempered by a sense of peace, knowing that we all stood together as followers of Jesus Christ, no matter what.

We were able to attend most of those Sunday evening services for several of the Vietnam War years. We will always remember Country Church as a unique blessing and privilege.

In the summer of 1973, we prepared to go on furlough. So much would be changed when we returned to Vietnam. As for us, our lives would never be the same again.

Broken Branch

Our family was looking forward to our upcoming furlough after being in Vietnam for our second term of six years. This was going to be a special furlough for us. One of our friends, an American Air Force major who had served in Vietnam had been transferred to Wiesbaden, Germany. His wife and son had joined him there and had kindly invited us to visit them. The plan was to take this perfect opportunity to also get to visit Switzerland before going on to Germany. We were all excited about that and also because, after that visit, we would be going on to visit our families and friends in America and Australia.

In April of that year, 1973, Janice (almost ten) and Jeanette (almost nine) were coming home to Danang, for summer vacation from the missionary children's boarding school which they attended for a couple of years. They were flown from Nha Trang to Danang on an Air America C46 (DC3). Oliver and I still hold in our memories a permanent picture of the two of them scrambling from the plane and rushing toward us and into our arms. We felt such pleasure being together again. After a short time, we all drove eighty-five miles from Danang to Quang Ngai, where we had temporarily rented a small house. Quang Ngai is where we were planning to hold our third teacher-training seminar for a group of twenty young tribal people we had recruited.

I remember one of the early seminars we held in Quang Ngai. I became sick with a severe chest cold, to which I am susceptible. It was cold and rainy, and I was feverish, hurting, and miserable.

Sometime in the wee hours of the morning, Oliver got up, dressed, and drove to the local pharmacist. He knocked on the door, woke up the pharmacist, and pleaded with him to produce something that would bring relief to his wife. He did. After taking a dose of some kind of medication, I finally got relief and was able to sleep.

In Quang Ngai as in Danang, the peace would often be interrupted with enemy rockets hitting the city. Sometimes we could hear them as they whistled overhead. We had to get down on the floor and under a bed or table. One time, the warning siren of incoming rockets sounded right after breakfast, and we all headed for the bunker on the property. Just as we got to the door, Oliver stopped short. Ngiah was trying to gently move him along from behind and urging him to go in the bunker. Oliver's momentary hesitation was because the bunker had several inches of water in it, and we Westerners don't relish walking in water while wearing shoes. Finally, we all got settled inside, but Ngiah still seemed puzzled over what he considered that odd hesitation at the door. He didn't ask, and we didn't want to talk about it, as he probably wouldn't understand. He might comprehend our not wanting to get our shoes wet but not about it being a problem when we were in mortal danger!

We had bought Ngiah a pair of new shoes, his first pair ever. This was pretty high class for him, who always either wore thongs or went barefooted. He valued those shoes, and we saw him more than once carry them under his arm in the rain so to not get them wet. Maybe we should have taken our shoes off and carried our shoes to the bunker, but then, unlike him, that would not have worked either, as that would have meant Oliver getting his clean feet into dirty water! Ngiah would never understand that to be a problem.

At the other two seminars, we had taught the teacher candidates to read their own Hrey tribal language. They could already read Vietnamese because they had attended some years of Vietnamese grammar school. Now, this seminar was to teach these same young tribal men and there was one woman, how to teach the materials we had prepared. We would train them to teach Hrey children elementary math, science, social studies, and to read and write their own language. We had prepared detailed teacher's guides and teaching charts for them and also our three reading primers the children would be using. The government had provided a schoolroom for these seminars and also board and room accommodations for the students.

Weather-wise it was hot, humid, and horrible. It was also somewhat tedious, as each of our students would be called one by one to the front of the non-air-conditioned schoolroom to practice teaching reading from our charts and writing on the blackboard under our ever-watchful eyes. Making good progress day by day as we taught and observed these students learning, practicing, and preparing to start teaching in their villages was encouraging.

In the evenings after our meal, we would spend time with our girls. I had brought a book I had gotten from my grandmother years before. We decided to read this together. I myself had not even read it. We settled down, and Oliver read to us. It was a lovely story of how a young lady in Bible school had to leave school because her mother had died and she needed to help care for her ten-year-old sister and be with her father. As she was leaving, the principal wisely told her that although she was leaving their Bible school, she was not leaving the school of God. The story brought out so many lovely Christian qualities. We were all enjoying it. But then the ten-year-old sister got ill and told her big sister that this must be "God's night school." Then the child died. That was so sad, but I was taken aback when Janice sobbed and sobbed inconsolably for some minutes.

We finished our teacher-training workshop with a closing ceremony given by the South Vietnamese Government Department of Education with lots of speeches. Then Oliver gave a speech in Vietnamese and handed out certificates and shook hands with our seminar graduates.

We had one more project to complete. The teachers would need blackboards, an important part of their teaching equipment. We learned that cement mixed with black paint and applied to a wooden board creates a blackboard-like surface. It worked beautifully. Since all of the students had been flown in by American helicopters from remote areas, they would go back the same way. Oliver measured the helicopter door, and the blackboards were cut to fit through the door diagonally. Finally, the newly trained teachers left for their respective villages with their teacher's guides, crucial teaching charts, blackboards, and plenty of chalk, which we obtained locally.

We were finished in Quang Ngai, so we packed up and headed back to our home in Danang. That trip turned out to be scary with our car slipping and sliding through a hurricane.

We finished final preparations for our furlough. We had relocated Ngiah and his family from Danang to Quang Ngai, where he would be holding Bible studies with his own people and especially Hrey soldiers in the area. Since it was not time for Pam Brady, our British coworker, to go on furlough, she would remain in Danang and carry on her Vietnamese ministry.

At breakfast on the morning we were to fly to Saigon to catch our international flight, as was our custom, we read from the *Daily Light* for that day, July 12. The verses spoke of the Lord going with us, which seemed meant for our family as we began our journey home. There was a poignant moment when we became aware of just how sad Janice was about leaving Pam behind. She got up from the table, got a Bible from a bookcase, and asked Daddy to help her find the verse she was thinking of. Then she looked up at Pam and read John 14:27: "Peace I leave with you, my peace I give unto you: not as the world gives, give I unto you. Let not your heart be troubled, neither let it be afraid." We then said our good-byes, and off we went to Saigon.

Upon arriving at the missionary guesthouse in Saigon, we learned that our exit visas were not ready as expected. Oliver then had to run around town to various offices to speed up the process so we could make our international flight for which we already had our tickets.

On July 13, the girls were playing outside. I was inside the house, putting the finishing touches on the Hrey hymnal that Ngiah and I had finished translating, which was going to be published.

The girls and I and other missionaries had eaten the evening meal when Oliver finally returned. I sat with him as he ate his late dinner and updated me on our visa situation. Suddenly, we heard a loud thud. We went to the door. Janice had fallen from a nearby tree and was lying on the ground, not breathing. We learned later that a branch had snapped under her foot as she was climbing down and that she fell only about ten feet, but unfortunately she hit her head hard on a circular curb around the tree. We carried her into the house and prayed immediately for her recovery, never considering that "No" is an answer too. As I looked around the room, everyone seemed frozen in position or moving in slow motion. Janice began breathing again but was still unconscious. The ambulance arrived, and in less than half an hour, we were at the American hospital.

Janice had a severe concussion. I sat with her all that night as she lay there so still and pale. She woke up once and said, "My head hurts, Mommy, it really hurts!" During the following hours, she kissed me once on the cheek and also kissed Oliver when he was with her.

Thirty-four hours after her fall, I was awakened before dawn by a knock on the guesthouse door. There was an urgent call for me to get to the hospital. My legs shaking, I dressed quickly, and our friend, Janie Voss, drove me to the hospital. Upon our arrival, Oliver told me Janice had died at 5:00 a.m. *Oh no, not my Janice!* my brain screamed. It was Sunday, July 15, 1973.

We returned to the guesthouse. I have a vivid memory of Oliver sitting on the bed with tears streaming down his cheeks and saying, "And now, we look to the future when we will meet Janice again." The ever-thoughtful guesthouse staff moved us to a bedroom in their new wing that they thought would be quieter.

We stood hugging and weeping in the center of that room. Then Oliver told me something he had completely forgotten. A month or maybe two before Janice's accident, he was returning to Danang from one of his many flights to visit the Hrey villages. He was not asleep, but he saw clearly we were standing and hugging each other and weeping because something terrible had happened to Janice. Aghast, he somehow purposely removed it from his mind. But the fact that this room looked exactly like the one he had seen, jogged his memory. In our agony, we pondered what good such knowledge would do now. Should we have somehow protected her? But how, and from what? It was not possible. What, then? We eventually were comforted by the firm realization that God knew this accident was going to happen, not that He had caused it. We also realized that He could have prevented it, but He didn't. That He knew it was going to happen was one thing, but that He wanted us to know that He knew was an inexplicable comfort to us.

We have often been asked, since He is God Almighty and claims to love us dearly, why did He let this happen to Janice? It is true that God loves everyone, but does that mean nothing bad will ever happen to anyone? For that matter, where is it written that God stops all pain, suffering, and death for Christians? The reality is, good and bad things happen to everyone.

Janice died on July 15, 1973, after having just turned ten on May 8. The funeral service was held at the International Christian Church in Saigon on July 19. The missionaries at the Wycliffe guesthouse were a tremendous blessing and help to us in so many practical ways during this time. I especially remember that one morning among the disarray of our toothbrushes, toothpaste, and other things on the shelf in our bathroom, someone had placed one beautiful rose in a glass vase. This blessed me deeply, more than words can express. Afterward, we went up on the flat roof of where we were staying and witnessed a beautiful sunset. It is impossible to express our sadness as we tearfully watched the sun disappear from sight.

The US Embassy kindly told us that it would arrange to take Janice's body to America, but we were concerned about how this might be interpreted by our tribal people. They feel compelled to have their dead buried near their home village because they are obsessed with the belief that the dead person will return and "bite" them if they are not. So we decided that our Janice would be buried in Vietnam, where she was born and where we served as missionaries, in the Christian cemetery in Saigon. Although none of us were in the mood for it, we continued with our planned trip to Switzerland and Germany on our way home to America and Australia.

When we arrived in Switzerland, we canceled the room that Jeanette would have shared with Janice. The hotel staff was so thoughtful and understanding. We were exhausted, so the three of us got into the big bed together, covered ourselves with the Swiss eiderdown quilt, and slipped into merciful sleep.

We had scheduled some sightseeing trips and tried to enjoy them. I remember the jolting moment when we went into a restaurant and asked for a table for three instead of four. Walking behind a child about ten years old with long blond hair just like Janice's was an emotional moment for me. Her arm was in a cast, and I thought, *If only, if only* . . . Most of our couple of days in Switzerland were foggy and rainy, which matched my mood perfectly.

We then flew to visit our dear friend Major David Silver and his family in Germany, which was difficult for all of us. Major Silver had known Janice in Vietnam. They were very kind and even had a nice early tenth birthday cake and gifts for Jeanette. But most of our time spent with them is pretty much a blur to me.

We flew to Minneapolis and were met at the airport by family and friends. We wept together. I will never forget my mother presenting me with a bouquet of ten beautiful pink rose buds in memory of Janice, our precious firstborn child and her granddaughter. The last time my mother saw Janice was when she was four years old as we boarded a passenger liner for the first leg of our return to Vietnam. Janice really cried then when she said good-bye to her dear grandma.

The months that followed dragged by and are beyond description. The pain from the loss was keenly felt hour by hour, day by day. Waking up each morning, my first thought was, *Janice is dead!* It seemed so incomprehensible. She was gone! Gone!

Oliver, Jeanette, and I tried comforting each other, which was a blessing. But in some ways, one must walk through the valley of the shadow all alone, yet sensing the love and prayers of family and friends. There is no doubt that the bitter became bittersweet as we found ourselves loved by so many people. Intermingled with all that was God's tender grace that we experienced but could neither trace nor explain.

> "How entered, by what secret stair;
> I know not, knowing only He was there."
>
> —T. E. Brown

Jeanette bravely went to public school alone. She was in the fourth grade. We were living at Bethany School of Missions. They graciously provided us with room and board. We were given the use of two bedrooms. They planned to paint the rooms and let us choose any color we wanted. Jeanette chose yellow to remind her of Janice being in heaven. A friend of ours gave her a set of yellow curtains and bedspread. Her room was bright and cheerful and did seem like a touch of heaven.

Life went on as it always does. We were given a room to use for our office in the basement of the administration building. We had our friends and family nearby and spoke at churches and to many small groups. People were so kind to us and were extremely interested in Vietnam and our work there. Life was busy.

Nevertheless, there were times when grief hung heavily on my heart. The whole hard, unexpected, shattering, bruising, wounding, hurting pain of Janice's sudden death sometimes would hit me all at once. And it hurt so much! Then one day as I was bowed low in grief before the Lord, I remembered a poem which had impressed me in Isobel Kuhn's book *Green Leaf in Drought-time*. It is by Gerald Massey.

> My cloud of battle dust may dim,
> And veil His splendor, curtain Him;
> And, in the midnight of my fear,
> I may not feel Him standing near.
> But as I lift mine eyes above.
> His banner over me is love.

Somehow the words of that poem brought the reality of His steadfast love into focus whether I felt it or not. By lifting my eyes above from that low point, I experienced a tremendous sense of His nearness, sweet relief, and welcome comfort. The pain did not go away completely. It was still there, like a big bruise that needed to heal. And bruises take a long time to heal. Occasionally, a word spoken or a memory would press on that bruise and the pain would return. Eventually, though it still hurt, sometimes it became okay. The hurt became commingled with gentle, inexplicable comfort.

I learned that it is not enough to know that God's grace is sufficient, but also that He does not force His grace upon us. His grace is like the sunshine that floods the room when you open the drapes and let it in. I had to open my heart and receive it by faith, not by feelings. Occasionally, tears still come, and yet through the dimness they cause, there is a beautiful light that sparkles with the realization of the glorious fact that our Janice is *home* and waiting for us, and one day we will all be together. That expectation is an indescribable comfort and is what makes the huge difference between the death of Christians and non-Christians in times of crushing grief. It is the supreme comforting light at the end of the dark tunnel.

When Janice was four years old, we had a calendar with a picture of Jesus on it that was a favorite of hers. She would look at it and say, "Oh, I love Jesus!" Sometimes as she grew older, she would say, "Jesus is a wonderful man!" Her little joke with her sister, Jeanette, was, "When Jesus comes, don't bump your head on the ceiling, Jeanette." This would be followed by peals of her exuberant laughter. Now that laughter is turned to pure joy. I keep a picture of her on our dresser, taken when she was eight. She is flashing her special smile, and her eyes are twinkling with happiness, reminding me that she is now in the presence of her beloved Jesus, joyful beyond description and waiting for us to join her there too.

When she was eight years old, we went to the Country Church service at the American Air Force base in Danang where they were showing a Christian film. When we got home, Janice told me she had cried in church, which I had noticed. She told me that she was not crying because of the movie but because, as she said, "I asked Jesus into my heart." Many will attest to the beautiful change that became more and more apparent in her young life as the months went by. Neither a child's beauty nor brilliance provides any real comfort when compared with the comfort from knowing that a deceased child had a personal, saving encounter with Jesus Christ.

There is more to the Janice story. On July 18, 1986, while ministering among Asian immigrants in California, we received an astonishing piece of mail from Pat Bonnell, a missionary friend of ours who had served in Vietnam but who was then serving in the Malaysian province of Sabah. Photos of an open grave were in the envelope. Next to the gaping hole in the ground was an open casket and inside it were the covered remains. A piece of paper was posted on one end of the casket, upon which "*THANH HOA*," Janice's Vietnamese name, was printed in bold letters. Our initial response was wordless astonishment! Squatting behind the box were five Vietnamese people, two of whom we knew and highly respected. One was the Reverend Doan Van Mien, the president of a large Christian church organization in Vietnam, and next to him was his wife. They all looked very sad as they faced the camera.

The letter informed us that everyone who had been interned in that Christian cemetery had been exhumed and moved somewhere in January 1985, but we had no idea where. One of Oliver's nephews, Bernard Trebilco, visited Vietnam and tried to find out where Janice had been moved but was unsuccessful. It really did not matter to us. But we were thankful to the dear Vietnamese Christians, who themselves were not having it easy just then, to have taken the time and trouble to take pictures and send them to our friend. They figured, and correctly so, that she would know where to send the pictures to us in California.

Would you believe that there is still more to the Janice story? Many years later, when Vietnam became more open, a Vietnamese immigrant friend of ours, Mr. Khoi, went to Vietnam for a short visit to his family. This Christian gentleman was always grateful for the years Oliver had patiently taught him Bible truths when he first came to America as a new Christian refugee and for many years after that. While in Vietnam on this visit, and unknown to us, he took upon himself the mission of finding Janice! And he did just that.

On February 26, 2007, Mr. Khoi found Janice's plot, number 126, in a Christian cemetery in Lai Thieu outside Saigon. He was not at all happy to discover that her tombstone, when translated in English, read, "Young man, Thanh Hoa," and the "Rest in peace" date was 1985, which was when she was moved there. Actually, Janice had been resting in peace since 1973.

When Mr. Khoi discovered these unacceptable inaccuracies he took immediate action. He phoned us from Vietnam to discuss the correct wording and made arrangements with a pastor friend of his in Vietnam to have the tombstone redone and insisted there must be a cross on it. He then returned to California and took care of all of the expenses. Quite a nice touch from the Lord through our friend, don't you think?

Many months later, we received an e-mail of a picture of the renewed tombstone with accurate information on plot number 126. The gray tombstone is three feet high and two feet wide with large bold, black letters on a white recessed background. Listed first is Janice's English name and then continues in Vietnamese, which when translated reads:

JANICE ALANE TREBILCO
THANH HOA
May 8, 1963 to July 15, 1973
WITH HER SAVIOR, THE LORD JESUS
WHOM SHE DEARLY LOVED
Daughter of missionaries:
Mr. and Mrs. OLIVER TREBILCO

It is wonderful to us to realize that our Janice has to move again. This is fine with us. We can hardly wait!

> For if we believe that Jesus died and rose again, even so God will bring with Him those who sleep in Jesus. For this we say to you by the word of the Lord, that we who are alive and remain until the coming of the Lord will by no means precede those who are asleep. For the Lord Himself will descend from heaven with a shout, with the voice of an archangel, and with the trumpet of God. And the dead in Christ will rise first. Then we who are alive and remain shall be caught up together with them in the clouds to meet the Lord in the air. And thus we shall always be with the Lord. Therefore comfort one another with these words.
>
> —1 Thessalonians 4:13-18

Three Dried-Up Prunes

I couldn't believe my ears. Three months after Janice died, Oliver told me that he thought we should have another child. With all the trouble I had experienced in childbirth, he had said, "No more children!" Now, however, he felt that the two of us with only Jeanette would be, as he put it, "Like three dried-up prunes." I was stunned! I was astonished! I was shaken! I was also forty!

Somehow, and I think somewhat strangely, the first thing that crossed my mind was that I would have to go through the process again of teaching a child to pick up his or her toys. Rather peculiar when I think of it now, but that must have epitomized what motherhood again would involve for me. In retrospect, however, it might make some sense because the simple task of teaching children to pick up their toys did involve a mom's weighty, constant obligation of teaching character qualities. After all, even such a simple task required obedience, endurance, stewardship, neatness, carefulness, responsibility, working together, thankfulness when helped, and the enjoyment of accomplishment. Actually, I have no idea why the toy thing came to mind. I might have also thought, "Oh no, not that again!" At any rate, for whatever reason, it crossed my mind for just a second.

When I recovered from the initial shock, I tentatively agreed but reminded him that surely we were hoping to return to the field as soon as possible, and pregnancy would certainly delay things. To complicate the situation, because of a medical problem, I had been advised by my doctor to have a hysterectomy, so when I told him of our desire to have another child, he just smiled and said it would probably never happen. My thought was that if after two months there was no child on the way, we should get back to the mission field. However, you do not dictate to God, because it was three months before Oliver got his wish, and we were expecting a baby. When my doctor said, "Yes, you are very pregnant!", it was my turn to smile. When I gave Oliver the news, he looked at me with awe, wonder, and delight and said, "I will give you up to half of my kingdom." We were both so thrilled.

Jonathan Oliver Trebilco was born on October 16, 1974, at 9:55 p.m. We did not give Jonathan a Vietnamese name because his name is in every Bible translation in all the languages of the world. He was named after Prince Jonathan in the Old Testament, the humble, fearless, honorable, valiant, selfless, God-fearing, faithful son of King Saul and the beloved friend of David. Jonathan is a Hebrew name that means "God's gracious gift." We dedicated him to the Lord as we had done with Janice and Jeanette.

He was born one day after my forty-first birthday. Never have I received a more fantastic birthday gift. He was wonderful, and we were so happy. Jeanette, now ten, was overjoyed. She sent a note to me in the hospital that had a picture she had drawn of four happy circle faces with the words, "Now we are four again!" Everyone was pleased, and those who had mourned with us now rejoiced with us. We were lavished with gifts galore at several wonderful baby showers for our son.

We waited the minimum time allowed to book a flight with a newborn, which was six weeks. Then we flew to visit family and friends in Australia before returning to Vietnam. Our fares were quite inexpensive because we agreed to stop for a couple of nights in Pango Pango, American Samoa, which included a stay at a rather nice hotel there.

Soon we were on our way again and finally arrived at Brisbane, Queensland. Much to our surprise, Oliver's estranged father was at the airport along with the rest of the family. It was there we learned that Oliver's mother had just passed away from a heart attack. They had sent a telegram, but we never received it. When Oliver learned about his mother's death from his brother, he went to hug his father. This time, they wept in each other's arms. So on Christmas Eve, Oliver had the sad task of being one of the pallbearers at his mother's funeral, whom he had not seen in nine years. The sorrow was even more poignant when we were told that she collapsed just as she had gotten up from a hammock to show someone a photo of her new grandson, Jonathan. We were so sad that she never got to see or hold him.

During part of this furlough, we stayed with Doreen, who was now married to David Weller. They were living with Dad Trebilco in Toowoomba, Queensland. It was hard to see Oliver's aging father, now a widower. Everyone was surprised that their mother had gone first since she was twelve years younger than their father. We had brought a sixtieth wedding anniversary gift for them, which produced more tears when we gave it to Dad Trebilco.

Oliver's dad was considered an expert at playing checkers, which Australians call "drafts." He almost always won and was a formidable opponent. One day, he invited me to play drafts with him, which I politely did although it is definitely not my thing. When we finished, he chuckled and had a twinkle in his eyes. I then realized that the rascal had obviously let me win!

We learned later that the death of our Janice and then the death of his wife had a tremendous impact on Dad Trebilco. Later, he confided to a friend that he was grieving over the realization that he would never see these loved ones again because he could not go to where they were. This was interesting since he was a self-confessed atheist and was not supposed to believe in a place called heaven. We then realized that his hard heart was beginning to soften. Seven years after the death of his wife, Dad Trebilco became a Christian at the age of 92.

On February 13, 1975, we returned to Danang Vietnam, with a joyful welcome back from Pam, Phuong, and Ngiah, who had come up from Quang Ngai for the occasion. The last time we had seen Ngiah was when Janice was still alive, and we were leaving for furlough. Now we had returned with our infant son. When we drove up to our house, Ngiah took Jonathan in his arms and gave thanks to

God for him. Phuong had prepared a feast, and we all rejoiced being together again.

Jeanette got right down to her school lessons, and Jonathan continued to thrive. Oliver made two road trips to Quang Ngai to encourage Ngiah and his family who were living there. They had started a small Bible study meeting for Hrey soldiers.

We were eager to take up where we had left off. Materials had been prepared, and now we were hoping to get more Hrey schools opened. We also planned to continue doing Bible translation, as we had only completed the gospel of Mark.

However, trouble was brewing. Communist forces were gaining ground. We kept track of the news and were once again deeply saddened to learn of more missionaries being captured. Every day found us referring to the map as province after province fell to the enemy.

One day, a big black car stopped at our gate. A lady from the US Embassy came to tell us that in case we needed to be evacuated, a helicopter would be sent to get us and would land on the empty lot at the end of our block. However, this plan was never implemented because as the fighting escalated, refugees from other cities and the countryside began pouring into Danang, so the lot was no longer empty. Never have we seen such a flood of tired, frightened people.

Every day, missionary men from various mission organizations in Danang met together to evaluate the situation. One of those days, I suddenly felt the urge to go to the closet, pull out our suitcases, and start packing. It was no surprise to me when Oliver returned from the meeting to report that everyone felt it was time to leave. We could not wait for the American officials to make arrangements, as we knew they knew if missionaries left their posts, the people would panic, knowing all was lost. They hoped we would stay as long as possible, which is exactly what we did. Now, though, we knew the time to leave had come.

On March 20, we began the painful process. We worked for five days destroying all records and financial statements lest any of our people be connected to us and falsely accused because of the ridiculous enemy lies accusing missionaries of being spies. We sorted out everything in the house. We gave all clothing except what we took on our backs or in our suitcases to relief, which Phuong took to grateful refugees huddled along our road in the once vacant lot and on the nearby beach.

On Tuesday, March 25, 1975, we left our house in the care of Phuong and were relieved that we were able to drive out of our property because our road was rapidly filling with refugees. We drove our station wagon to another mission property and left it there. Then we went with other missionaries to the Air America terminal. It was crowded, but eventually we boarded a small plane and were flown to Saigon. Five days later, on Easter Sunday 1975, Danang fell to the enemy, and the Communist flag went up. Sadly, Danang now belonged to Communist North Vietnam.

Many years later, we learned from Phuong that our house was completely ransacked and looted, not by enemy soldiers, but by hard-core thieves. A truck drove up to the house after we left, and then another truck drove up. The thieves in one of the trucks killed the thieves in the other truck, and Phuong was in grave danger. She was threatened with a gun to her head. She was finally able to flee to some Christian friends who lived nearby. At that time, she had her little sister with her and also a young Hrey girl, an orphan, for whom she was caring. They escaped with her.

When we touched down at Saigon, the lady on the plane to whom I had given a baby blanket to cover three infant orphans in a basket that were in her care handed it back to me. Without thinking in the rush of things, I took it as we were making our way off the plane. We knew that the next thing we had to do was get a flight out of Saigon as soon as possible. However, we soon discovered that it was not going to be that easy. We were mercifully unaware that soon we would face the possibility of becoming a family of only three again, like "three dried-up prunes."

Panic in Saigon

The missionary guesthouse was full, so we booked into a hotel where there were a number of other missionaries from several different organizations from around the country also getting ready to leave. Every Vietnamese person we met in Saigon seemed to be in denial that South Vietnam was on the verge of losing the war. Accepting the imminent fall of the country into the hands of Communist North Vietnam was too painful. Many of these same people had fled from the North in 1954 to escape communism, so they already knew what suffering and hardship to expect. I felt a twinge of guilt, realizing that we could hopefully fly away to freedom. But I also knew that staying would only endanger others. Pam had already gone on her scheduled furlough to England, and Ngiah, E-Ne, and their four small children were trying to make their way back to their village at Bato. In Saigon, our most important task was to book a flight out of country. It did not seem real to me that this was actually happening. Vietnam had been our home for fifteen years.

Soon after our arrival at the hotel, Jonathan, now five months old, became seriously ill with severe diarrhea, a high fever, and was vomiting. A missionary doctor at the hotel gave us antibiotics to give to Jonathan, which were not working. He was pretty sure Jonathan had viral gastroenteritis, which was later confirmed. We surmised that the baby blanket I had given the woman on the flight from Danang and which she had given back to me had been contaminated. Jonathan's condition worsened and was considered very serious, so the doctor advised us to take him to the American hospital in Saigon.

A missionary nurse also staying at the hotel told us later that she did not expect our son to live.

After Jonathan was examined at the hospital and admitted, I was instructed to give him a dose of a red, oily liquid at prescribed times. Every time I gave it to him, though, he gagged and choked, and I had to throw him over my shoulder. I felt so helpless. I was in a ward with half a dozen other concerned mothers, caring for their sick children. The awkward thing for me was the typical Vietnamese reaction of a little nervous laugh from the mothers every time Jonathan choked!

I sat by him as he lay in a crib throughout the long night. I felt calm but also concerned. Toward morning, I could see he was sinking. He looked at me with pitiful, sunken eyes. I had been praying, for hours. Suddenly, I prayed, "Lord, send me an angel!" I do not know why, nor have I ever prayed such a prayer before or since.

I went to the desk and called Oliver at the hotel. He in turn contacted the missionary doctor there. When that doctor learned that Jonathan was not being rehydrated intravenously, he became upset, grabbed a cab, and took off for the hospital. He went to the desk, identified himself, and asked for the necessary intravenous equipment. The nurse at the desk said she would first contact Jonathan's doctor. Our doctor friend banged the desk with his fist and commanded, "First, you will give me the equipment, and then you may call his doctor."

Oliver had arrived, and someone took me back to the guesthouse to try to get some rest. An hour passed before the missionary doctor was finally able to get into one of Jonathan's veins by doing a cut-down on his ankle. We are convinced he saved Jonathan's life. That doctor was not an angel, of course, but he sure seemed like one to me. After only a few hours, I was told that the improvement in our son was extremely encouraging. He spent another day in the hospital, and then it was worked out that we would pick him up from there on our way to the airport, as Oliver had been able to get new flight reservations after four cancelations and rescheduling four times.

However, when we picked up Jonathan at the hospital on April 1, he still did not look good. Oliver laid him on a pillow I had on my lap. We got to the airport, checked our baggage, and waited for our flight to Singapore. There were several other missionary friends from various organizations also waiting for that flight.

Suddenly, Jonathan started having convulsions. We were in a panic. Stan Smith called for an ambulance, Fred Donner ran around the airport collecting all the ice he could get his hands on to help bring Jonathan's temperature down, and Janie Voss, who was Jeanette's boarding school teacher, took charge of Jeanette. Others were also so kind, all angels in disguise, no doubt.

We were rushed to the hospital in the ambulance with Jonathan. The doctor in emergency there seemed to dilly-dally, and Oliver was in no mood for that. He told the on-duty American doctor that he had already lost a daughter in this hospital and was not about to lose a son. We learned that Jonathan was having a reaction to a medication he had unfortunately been given just before we had picked him up on our way to the airport. After the doctor gave him an injection of the proper antidote, we went to the missionary guesthouse, where there was now room for us. The hallways were lined with clothing and other items that were being discarded, as everyone at the guesthouse was packing to leave. Saigon was in turmoil.

That night, Oliver returned to the airport to get our suitcases that had gone to Singapore and happily had been returned to Saigon. While all this was happening, South Vietnam was being overrun by enemy forces. Finally, we were able to get new reservations for a flight on April 3, 1975, leaving Saigon for Singapore. By that time, the situation in Saigon was very tense. We learned later that our plane had actually been ordered to return to Singapore because of enemy activity at the Saigon airport. Then at the last minute, the order was reversed and the plane turned back. Soon we were all settled in our seats and prepared to leave.

I cannot adequately describe my feelings as that plane lifted and I saw Vietnam gradually disappearing from sight. We had often smiled when American soldiers' tour of duty was finished and they quipped that they were leaving with mixed emotions: "Joy and gladness!" For us, however, there was grateful joy because we were all safe but also awful sadness because we were leaving. There was a kind of numb disbelief.

As I held my weak but recovering son on my lap and with Oliver and Jeanette nearby, I felt a keen sense of relief and gratefulness to the Lord. Yes, we were safe, but our hearts were heavy as we wondered about the safety of our national friends and coworkers left behind in Vietnam. It was a dark time for the Christians, both Vietnamese and tribal. We felt unspeakably sad because we were sure many would, and we know did, suffer severely. We ourselves had lost a lot: Two languages and our literacy, our teacher training, our translation, our teaching and preaching ministries, our many friends, and almost all of our possessions. Yet we knew that this was nothing compared to what those left behind would be forced to suffer. We knew they would be sorely tested. We also felt with much sorrow that the time for foreign missionaries as we knew it in Vietnam was over. Missionaries had to leave, but God did not leave Vietnam. We experienced peace, realizing that He would be with His people through all that lay ahead. Christ's immutable promise was for them and for us too: "Lo, I am with you always" (Matthew 28:20b).

I learned a huge lesson from all of this. During our time in Vietnam, we were given a window of opportunity to accomplish certain things, but that period of time was temporary and had ended so abruptly. A motto that had profoundly impacted me in my youth was, "Live with eternity's values in view." I pondered the all-important *now* of life. Tomorrow may never be given, yesterday is beyond our control. Now is the time to take advantage of every opportunity of service while it is available. Some things gradually come to an end. For us, this end was instantaneous. We had absolutely no idea what lay ahead for us, and we experienced awful heaviness, grief, and loss on that trip as the plane took us away from Vietnam and on to Singapore.

In Singapore, we stayed at a hotel for awhile, and then an American family there on business generously opened their home to us. On April 30, 1975, newspaper headlines read, SAIGON SURRENDERS. It was over! Yes, and it was over for us too. Now we needed new direction and knew it would eventually become clear what that direction would be.

Nha Trang, Vietnam

Literacy Workshop resumes in Nha Trang

Developing Hrey educational materials

Cyclo: Door-to-door service

Our family "car"

Danang, Vietnam

Our "New" Scout Jeep
Rex Cougar (in middle) Earl Kilpatrick (on right)

We Love It!

E-Ne Helping

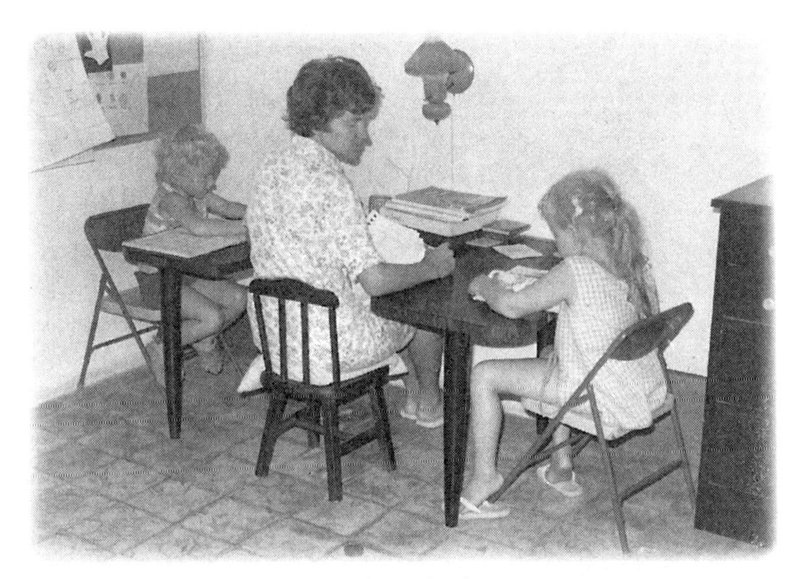

Home school

Hoyt Richardson, reading primer artist

Bottom: Newly printed primers arrive

Teacher Training class

Student teacher practice.

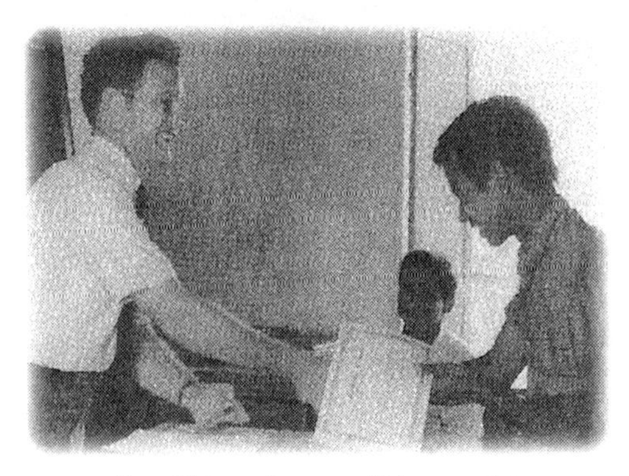

Certificate of accomplishment day

Our family—Vietnam

Saigon (Ho Chi Minh City), Vietnam

On to Indonesia

There had been changes in our mission arrangement in Vietnam. We had for some years been sponsored by Bethany Missions, who now agreed that we should join their work in Indonesia. There was a lot of government red tape involved and even a trip back and forth from Singapore, but we were finally granted permission to live and work in Indonesia.

Strange how we think our lives will turn out. Oliver and I both thought when we went to Vietnam that we would live and die there. Suddenly, everything had changed, and we found ourselves in a whole new world, on the island of Java in Indonesia. First of all, the only way we could get into Indonesia was because Oliver was allowed in as a *pendeta* (pastor). Never could the word *missionary* be used on permission papers to enter or remain in that country since that presupposed you were on a mission to convert people to Christianity. We were allowed into the country to preach and to teach Christianity to Christians but not to share the Christian gospel with anyone of another religion. As for me, I was allowed in as *ikut suami* (following husband)!

Now we began anew the interesting yet arduous task of learning another language—number three after Vietnamese and Hrey. Again, we were so thankful and amazed at how our prayer partners and supporters faithfully prayed, supported, and stood with us in this new assignment. Indonesian is polysyllabic with words so long that it's been said to take about one-fourth more pages to translate any text from English to Indonesian. For me, a big difference in this new language learning task was that this time I had two children, and one of them was not even a year old. Also homeschooling continued for Jeanette, who was now eleven.

We made every effort to master the language, but there was sometimes confusion in this task because some words were similar to the Vietnamese language. For example, the word for five in Vietnamese was almost the same as the word for six in Indonesian. The word *di* in one language meant "to go," while in the other language, *di* meant "to be at." Talk about confusing! But we pressed on. Gradually, Indonesian came more to the forefront, and Vietnamese and Hrey were no longer on the tip of our tongues. We were getting along quite well.

One day, some excited Christian ladies came to the house and informed us that a fishing boat of Vietnamese people who had fled Vietnam had just landed nearby. These ladies were providing them with food but wanted us to go and speak to them.

We arrived at the site and found ladies serving food to thirty-eight Vietnamese, including men, women, and children. I still vividly remember seeing one of the ladies with a large, blackened teakettle cheerfully pouring tea into glasses as she went around serving the grateful group. We felt somewhat rusty in Vietnamese, so we thought it best for a spokesman from the group to tell us their story, which he did. By then, our Vietnamese had "kicked in," and we had a wonderful time with these brave people who had fled Vietnam, risking their lives and the lives of their children to find freedom, especially freedom of religion, rather than live under atheistic communism.

Another time, these same ladies came to get us, but this time it was a ship full of Vietnamese who had fled from Vietnam. There was, however, a severe problem among them. At least half of the passengers were on the ship because they had somehow deceived the captain by getting on board in Vietnam themselves rather than the friends of the captain who had paid him but had been left behind. Now these passengers refused to return to the ship because they said they feared for their lives.

The ship was on its way to Australia. We spoke to the captain and warned him that we were going to notify Australian immigration and give them the name of the ship and the exact number of the people on it. Oliver made that call, which calmed everyone down. Finally, they left and all arrived in Australia safely.

Ironically, when we were in Sydney some months later, we were asked to visit a housing complex that the Australian government had provided for Vietnamese refugees. Just as we arrived, we saw a medical emergency vehicle rushing to help someone who had collapsed in one of the apartments. When we got to that apartment, we discovered that the man who was ill was the husband of one of the frightened ladies we knew from the ship that had stopped in Indonesia. We had a grand reunion with many of the folks from that ship, which included a feast, of course. We were glad to see them all safe and sound in Australia.

We spent two years with Bethany Missions at Ungaran, Java, in Indonesian language study and to orient ourselves in this new-to-us country. There were more adjustments than we imagined. Most predominately was that in Vietnam we were highly respected and even honored as missionaries, whereas in Indonesia, we were considered to be infidels, people without religion, or at least without the "right" religion.

After two years of intense language learning, we moved to the city of Surabaya where we rejoined WEC International. Oliver was busy teaching evening Bible study classes in various homes.

At this time, Jonathan loved play-acting certain characters, and I enjoyed creating proper outfits of these people. He was young David for awhile with his sling. Then he wanted to be Goliath with a beard and a spear. Once, he wanted to be a Roman soldier. The outfits were very simple, very improvised, and never elaborate. They took the minimum of time to create but gave Jonathan the maximum enjoyment.

One day, Jonathan wanted me to make him a Lambretta! Somehow using a lot of cardboard, paper, and tape, I managed to transform more or less his tricycle into what might pass as a Lambretta. He was elated and was especially impressed by the windshield I had made with clear plastic. In his make-believe world, Jonathan made Lambretta trips as the driver to many places but actually never left our yard. He had a problem though when he wanted to be the passenger and alas, there was no driver. I left that one for him to solve.

In 1979 after a furlough, WEC assigned us to work on the island of Madura, which is just across the Madura Strait from Surabaya and is reached by ferry. We made our first trip to look for a house to rent. We found an excellent place that had a large, L-shaped living room, a large dining room, and four bedrooms. It also had four small rooms in an adjoining building, each with a typical squat toilet. We decided to rent it when the owner of the house promised to put a Western toilet in one of those rooms.

There was another small room with an inbuilt tile enclosure to fill with water for splash baths. Everyone takes splash baths in Indonesia. All that is required is water and a dipper. We never took to taking splash baths. Happily, there was another small tiled room in the main house where Oliver put in a shower. We were relieved that the house was surrounded by a fence and had a lockable large front gate.

The day we arrived to move into that house, an Indonesian-speaking Madurese woman was there to meet us. She had been a servant for a former missionary family some years before and wanted to work for us. We were so pleased! She was a wonderful worker and a good cook but was not a live-in servant. One thing was really different to me: She insisted on washing the tile floors of the living room and dining room every morning before breakfast! Apparently, this was something learned during the three hundred years that Indonesia was a Dutch colony.

Just before he was five, Jonathan was eager to learn to read. He made such good progress, and how he loved it. One night when he was six, he was in bed reading his Children's Bible. We were doing something in the hall nearby when we heard, "Some guy sure loves some girl!" When we asked him what he was reading, he said, "Song of Solomon"!

We loved our Madurese servant, and because of her, we first learned how much our Jonathan enjoyed preaching. We had many student preachers in our Sunday meetings, and each one had his own style. One might bang the pulpit, another might put his hands in his pockets, and another might walk back and forth. Jonathan, now six years old, would preach to our servant quite often, using each one of these point-making methods. She sat carefully listening to him preaching his heart out to her in Indonesian. I actually had to tell Jonathan that he needed to get permission from me to preach because our servant had certain duties to fulfill. How that boy loved to preach! Today, Jonathan is an ordained minister of the gospel and still loves to preach. The biblical Prince Jonathan was an expert swordsman, and our Jonathan skillfully wields the sword of the Spirit, which is the Word of God.

We were so thrilled when, while living on Madura, we were able to buy a new blue station wagon with the money Oliver had inherited from his dad, who died just as we began working on Madura. We purchased a new Chevrolet "Luv Truck" (pickup) from a dealer in Surabaya. The standard procedure was that they would then convert it into a four-door station wagon. When it was finished, it looked like it had come like that from the dealer.

We did most of our shopping for food and supplies in Surabaya once a month, and traveling there in that station wagon with air-conditioning was a real blessing. Before the car was finished, we had traveled by bus to Surabaya, which was crowded and not air-conditioned. We were not allowed to open any of the bus windows, as the other passengers said the sun would be hotter with the windows open! We survived, but you can imagine how much we appreciated our new vehicle. The whole trip each month now was more like a pleasant family outing.

The fresh fruits and vegetables available in Surabaya were fantastic. There were juicy, sweet pineapples, huge papayas, large mangos, wonderful bananas, and the list goes on. We could also buy all sorts of canned foods and necessary supplies. On shopping days, we would eat at a restaurant. Almost all the meat in Indonesian restaurants is prepared extremely spicy, so much so that we could hardly recognize whether it was chicken, lamb, or beef. We usually decided to eat at a Chinese restaurant, which solved our problem.

We all fell in love, though, with Indonesian chicken, or goat *satay*. They make this by stringing marinated chunks of meat on bamboo skewers, grilling them to juicy perfection, and serving them with their magnificent peanut sauce and steamed white rice. We usually had *satay* at home once a week, on Tuesdays. It was quite a project to prepare the meat and the sauce required peanuts being crushed to smoothness by hand with a pestle and mortar.

One Tuesday, we had run out of chicken, so I gave our servant instructions to prepare something else. Unknown to me, she slipped off to the local market and bought a chicken. When she served that *satay* meal, her smile of victory, the twinkle in her eyes, but mostly Jonathan's delight were so special. There was no way she was going to disappoint Jonathan. I think *satay* and peanut sauce will always be one of his favorite foods.

We also enjoyed the church picnics when each person was handed their meal. Opening the carefully wrapped banana leaf, we would enjoy rice, slices of tender beef in a delicious sauce with some vegetables. Chili peppers were served separately. *Ketcup manis* (sweet sauce) is a sweet molasses-like sauce, definitely delicious and definitely Indonesian. Our word *ketchup* (tomato sauce) is an Indonesian word, which means "a sauce for meat."

Indonesia is right on the equator, and I discovered that I could set my watch by sunup and sundown, which was 6:00 a.m. or 6:00 p.m., respectively. Something that always intrigued me was that every day at dusk, the bats would leave the big mango tree in front of our house, and then the birds would arrive to roost for the night.

There were certain important cultural things we had to learn. For example, it was never proper to walk fast, especially for a woman. Proper decorum dictated a slow, got-all-day kind of walk. Also, it was rude to receive or give anything with your left hand, which was hard to remember. Once, we designed a Christmas card with the greeting written in red, which was a mistake. We learned that in Indonesia, red is the color of anger. However, in Vietnam, red is the color of happiness. We also learned to say "permissi" when passing anyone on the sidewalk, which was asking for permission to do so. Another very different thing for me was to learn that one never opened a gift in front of the giver. When I asked why, one person told me they did it that way just in case the person did not like the gift.

One afternoon a Maduresе gentleman came to visit us. I forgot that it was Ramadan, their fast month when they fast all day and eat all night, and I served him cookies and tea. When I remembered and apologized, he was not concerned and said that my forgetting was a blessing. I am not sure, but I think he meant that he was supposed to be fasting but felt he had to partake to be polite so therefore it was a blessing to him. Or was it a blessing to me because he did not embarrass me? I decided not to question him about his meaning.

Something we never learned to appreciate or get accustomed to was the loud, clear call to prayer six times a day issuing forth in full volume from a loudspeaker at the local minaret: predawn, dawn, noon, afternoon, evening, and night. I heard a lady say once that to her it sounded like the beautiful song of a bird. Not to us!

Jeanette kept busy in home school. We learned later she also kept busy many nights, as there was a pond outside her bedroom window. The croaking frogs would wake her up, and unknown to us, she would get up, slip outside, grab the frogs one by one, and throw them over the fence into our neighbors yard!

Jeanette had several friends among the church youth. One girl especially would pick her up on her motorbike and take her to her house on Friday evenings, which was a nice change for Jeanette. If we thought Vietnam was without outside entertainment, Indonesia had even less. We were happy that Jeanette loved to read which kept her occupied. She gave Jonathan a weekly Sunday school lesson and did many projects with him.

I felt it important that, besides ministry and homeschooling, I should try to provide interesting pastimes. Jonathan developed an interest in acting out stories by using the cardboard characters I had made for him to use in his play-acting, which kept him busy for hours. He loved Bible characters and super heroes. My characters were not perfect, but he seemed to think so. He had plenty of storybooks with pictures that he referred to for "accuracy"! There was one time when he was not satisfied because one of the Bible characters, Samson, was not quite right. I had copied and enlarged him from the shiny cover of a book, which I thought was a nice cardboard cutout figure. But alas, it was not glossy. Eventually, Jonathan did approve of it after I covered Samson with shiny plastic wrap.

One day, I was able to talk Oliver into wearing some things that would, with a lot of imagination, pass as a Batman outfit with the finishing touch of a construction paper half-mask with pointed ears. When Oliver jumped out from somewhere, Jonathan could hardly believe his eyes! He was first shocked but then impressed. Oliver is definitely not into dramatics, which made it even more hilarious. Jeanette and I became weak with laughter at such an unusual display of father-son togetherness.

Back in the days when we were still in Vietnam, and Janice and Jeanette were seven and six, I began teaching them to cross-stitch by drawing simple patterns on squares of cloth with large *X*s for the stitches. On Madura, Jeanette, then sixteen, became very adept at crewel embroidery. Every year, I would design a cloth wall-hanging depicting some aspect of the Christmas story, and she would then apply her embroidery magic and turn it into a colorful work of art. One of these was sent to Grandma and Granddad in America every year. They are still part of my valued Christmas decorations.

One day, we got a visit from an English lady who had just moved to Madura with her husband. He worked for a large international company. She informed us that the only reason she was willing to come to Madura was if the company her husband worked for agreed to put in a swimming pool at their house. She had come to invite us over to swim whenever we wanted. That was a tremendous blessing to our family and a fantastic luxury. Our friend and her husband also invited us to their large company Christmas dinner. We thoroughly enjoyed the delicious traditional British Christmas meal.

There were other unexpected treats for us on Madura. Oliver traveled every week twenty miles on the back of Pastor Wiyono's motorbike to preach in a church at Ketapang on the northern coast of the island. He would bring home large bags of delicious shelled cashew nuts.

Sometimes people seem to think missionaries and especially their children are deprived. That is not how it was for us. We have heard of parents who take their children to foreign places to expose them to other cultures for enrichment and educational purposes, and we did that and much more in our ministry travels. Granted, there were a lot of adjustments to make. Living on the equator without air-conditioning might be a form of deprivation, but living there and not just visiting other cultures was a tremendously enriching experience for all of us.

Also, I should mention, in every place where I lived overseas, as was the norm, I had a servant who did the washing, ironing, cleaning, and cooking. If that's deprivation, bring it on! Keep in mind that in the United States we also have servants, so to speak. For instance, in our country if we are having chicken for dinner, we do not have to buy a live one and kill and pluck it before cooking. Then too there is our dishwasher "servant," a real helper. Our automatic washer and dryer "servants" are always available for us and ready to work with just the push of a button.

Southeast Asia has made its mark on our family. All of us are now quite content to eat American, Australian, German, French, Swedish, Russian, Greek, Mexican, Italian, or almost any ethnic food. Yet, every once in awhile, we just have to have a bowl of steamed white rice with cashew chicken, or crispy fried spring rolls with fish sauce, or spring rolls with peanut sauce, or *satay* with peanut sauce, or a stir-fry dish, just so long as it is Asian cuisine.

We love the people we worked with in Vietnam and Indonesia and learned to appreciate many of their customs and worldview. We were immensely blessed by their friendship.

Our two years on Madura, though, was coming to an end. This time, we would not be abruptly evacuated out of our ministry as had happened to us in Vietnam. The plan from the beginning was to leave a growing church with capable nationals in leadership who no longer felt the need to have a resident missionary. We had literally worked ourselves out of a ministry. Now we would have to move on. Move on to what, we wondered? We had no idea. Yet I can honestly say that we had no concern because we had learned that our lives were about taking one God-guided step at a time. At the moment, we didn't know what that step would be but were confident we would soon find out. And so we did.

Two Houses

Oliver had been teaching Theological Education by Extension (TEE) in Indonesian for two years from mid 1979-1981, in the city of Pamakasan on the southern side of the island of Madura in the country of Indonesia. My contribution was offering hospitality to visitors, teaching women and children, coordinating a DVBS program that was a happy success, and creating costumes for the dramatization of the Christmas story. Once, Jeanette and I put on a simple, thought-provoking skit for the ladies of the church. We planned, we practiced, and we performed.

Gradually, we were seeing young men taking on leadership roles in the church. God was doing wonderful things. Soon it would be time for us to go on furlough, as we had served in Indonesia for six years.

We were all praying that we could build a church building, but getting permission was extremely difficult and was not happening. The church meetings were held in our rented house in the large living room. The attendance was excellent, mostly young, unmarried adults. The benches were stored on our front veranda between meetings and were replaced by our ping-pong table, which brought a lot of fun and entertainment for our family and others.

Oliver and Jeanette love the game, probably because they are so are good at it. I am an excellent onlooker. Sometimes Oliver played with a Christian neighbor, an ethnic Chinese gentleman, who was also very good at ping-pong. So he and Oliver were well matched.

Oliver's students would gather around that same ping-pong table weekly for his TEE lecture class.

We definitely needed to find a place for this congregation to meet after we left. The congregation was made up of ethnic Chinese Indonesian citizens. Having been born there, they could speak both Indonesian and Madurese. Anyone on Madura of Chinese descent was allowed to meet as Christians or could be evangelized, but evangelizing ethnic Madurese was absolutely forbidden by the authorities.

One day, we decided to challenge the church about the need for a church building. We started a 5:00 a.m. prayer meeting to pray about this urgent need. A few joined us, and oh, how we all prayed. One day, several of the church leaders came over from Java, to inspect some property that might be available. We all went to see it, and it was dutifully measured but did not work out.

Finally, someone found a house for sale that had been used for automobile repair. It was a dirty, sad house. Nevertheless, being somewhat desperate, we were thrilled that at least something had finally been found. After prayer and proper discussion, everyone agreed that this was "it"! Soon, all was settled with a down payment that the church had been collecting for some time.

Next was the cleanup process. It was a lot of hard work, but what an accomplishment when that dark, dirty place was transformed into a house of light. The benches were moved from our veranda and put in place. What a day of rejoicing when we finally began meeting there. Now we felt free to leave for our furlough with the confidence that God was doing good things in their midst.

In Indonesia, I learned to be careful as a woman in presenting any of my ideas or suggestions to the church committee. If I had an idea, opinion, or suggestion, I asked Oliver to present it to the church leaders. Whenever something was accepted and implemented, I was glad, but to be honest, I sometimes felt left out, no doubt because I was a woman from the West and not from the East. At any rate, it was somewhat hard for me. In Vietnam, Oliver and I had been co-translators and worked closely together on everything.

I will never forget, though, the Madura church's farewell send-off for us. It was a nice meeting with many expressions of gratefulness, mostly for Oliver. They were immensely grateful to him, their TEE teacher. I was truly happy for their gratitude for his tireless service.

After the meeting, they loaded us with many gifts, which in that culture was not appropriate to open until we got home. I was shocked that among those gifts was a small gold, diamond ring for me. In fact, most of the gifts were for me. Imagine that? The lesson was that working behind the scenes was not unnoticed and in this case produced unexpected gifts. I was heartened.

We were sad to have to say good-bye to these dear people and drive some twenty miles to Bangkalan to catch the ferry, which would take us across the Madura Strait to Java to catch our international flight to Australia.

Furlough was always a challenging and busy time traveling, speaking at meetings, sharing with prayer partners, and being with friends and family after so many years apart. Our excitement was somewhat muted, though, as we were concerned about the Madura church's debt. When we arrived at the airport in Sydney on July 22, 1981, we were amazed to learn that our mission had certain funds that they routinely distributed on a rotating system to the fields. It was Indonesia's turn, and our mission decided to send the funds for the Madura need. This was perfect. We and our friends in Pamakasan were thrilled and encouraged.

After two months in Oliver's Australia, we flew to my Minneapolis. We had to get there in time for school in September with Jeanette being a senior and Jonathan going into the second grade. Before leaving Madura, we shared with Jeanette and Jonathan that we felt the Lord was telling us He would provide a house in Minneapolis for us to rent during our furlough. "Oh boy, that will be great!" exclaimed our son. "I want my bedroom to be upstairs." The last time we were on furlough we lived with my sister Kathy and her husband, and once again, they had graciously invited us to share their home with us. Their generous spirit touched us, yet we felt that the Lord would give us a place of our own.

Jonathan's seven-year-old enthusiasm amused us, but we were all pretty interested in seeing what the Lord would provide. In the coming weeks, we often prayed about the house and sometimes even thanked Him in advance for what we felt He would provide. Jeanette and I sometimes talked about "our" house and wondered what it would be like.

When I wrote to my mother explaining our idea, she and Kathy began looking around for what was available. There were places for rent, but they were much too expensive for us. Our rent on Madura was $100 a month. Weeks passed and still nothing suitable had turned up. Time was running out, as we expected to arrive in Minneapolis early in September of 1981.

On August 30, we landed in Honolulu and called my mother in Minneapolis. She was so excited because, just the day before, a friend from church told her that she had heard that someone's sister was moving into a retirement facility and wanted to sell her house as quickly as possible. Our friend then asked if they would consider renting the house to missionaries for a year. The answer was a positive yes, except there already was an interested buyer. That would-be buyer, however, changed his mind, and the house was ours to rent and was extremely inexpensive.

A few days after our arrival in Minneapolis, we went to see "our" house on 29th Avenue South and were able to check out the downstairs. The renters upstairs had been given a month's notice and would be out by October 1. It was a joy meeting the owner, a dear elderly lady who loved our Lord and had always had a warm place in her heart for missions and for missionaries. The house was right, the timing was right, and the price was right!

We moved into the house and had the entire two-bedroom duplex to ourselves. The upstairs kitchen became Oliver's office (he has never had so much storage space before or since), the upstairs dining room became our den, and the one upstairs bedroom was for Jonathan. Oliver and I had the upstairs living room for our bedroom. Alas, it had no closet, but I found a large cardboard collapsible closet stored in one of the hall closets. Perfect! Jeanette was happy to have the one downstairs bedroom with her own private bathroom. The downstairs living room, dining room, and kitchen completed our cozy nest.

In the following weeks, we saw the Lord provide for us in so many ways. Over and over again, we were able to buy what we needed at garage sales and received furniture and other household items from family and friends. Some things were on loan; some things were given. We felt the Lord's love and provision flowing quickly through His people to us, and our hearts filled with gratitude.

It was wonderful to sense the Lord's obvious help in so many details. Once at a garage sale, we bought two twin mattresses, but they only had one twin bed frame. Soon after that, a couple from our church told us they had a twin bed frame we could use, but they didn't have a mattress. How we rejoiced together when they learned we only needed the frame! Again and again, the Lord led in remarkable ways to help us get things we needed to live for a year in the house He had chosen for us. It was a wonderful house for us, and we all loved it. In that house, we held a large high school graduation party for Jeanette.

Tucking our seven-year-old Jonathan into his snug, warm twin bed in his upstairs bedroom our first night in that house was a special moment as we pondered again God's provision for us. How Jonathan loved his room. He loved being upstairs, which was not guaranteed when we first talked about "our" furlough house back on the island of Madura those many weeks ago. He loved the red rug. He loved everything and declared it is the best house in the whole world. With a big smile and shining eyes, he said, "It is so great!" Yes, little guy, you are right. "The Lord hath done great things for us, whereof we are glad" (Psalm 126:3). He had graciously provided two houses, the House of the Lord in Pamekasan, Indonesia, and now ours.

We planned to spend a year in Minneapolis, and then we were not sure what was our next step. We did know that on Madura we had worked ourselves out of a job, as faithful, trained men stepped into leadership roles there. We wondered what lay ahead but were confident that God would open up new avenues of ministry for us.

That was certainly a year of changes. My mother went to be with her Lord. It was good to be home at that time but also sad.

Jeanette graduated from high school with honors and then went on to Wheaton College in Chicago, a big change for us. We all missed her so much.

After much consultation with WEC, it was decided that we should be assigned to work among Vietnamese refugees in California. Once again, we would be moving. But this time, we were able to take some of our possessions with us. Among the most valued was my mom's refrigerator and her double-oven gas stove. After some research, we discovered that it would be more economical to hire a section of a moving van to take these things plus some of our acquired furniture to California rather than start out with nothing.

We said our farewells to family and friends and headed west to start all over again. We clearly knew that this next step in our journey was God-guided. We were on our way, and the most special thing of all was that we would be meeting brand-new Vietnamese refugees who could not yet speak English. All those years of learning that difficult, six-tonal language were not in vain. Seven years after we had left Vietnam, we were on our way to once more be among Vietnamese people. But this time, they had come to us. Certainly this was not what we would or could have anticipated, but we were excited about the opportunity. California, here we come!

Asian Refugees

We had purchased a new burgundy station wagon with the money we had received from the sale of our blue station wagon to another missionary in Indonesia. In the autumn of 1982, we began the long road trip to Sacramento, California, from Minneapolis, Minnesota. Preparing to work as missionaries to the Vietnamese in the United States rather than overseas was a strange and new experience for us.

Our west coast WEC representatives welcomed us into their home. Just to keep things interesting, Jonathan developed chicken pox the day we arrived. We began our search for a house to rent. I thought we would easily find one by checking the want ads in the local newspaper. However, we discovered that things had changed, and that the best way to get results was to pay an agency that gave us a listing of houses that we might be interested in at our choice of location within a certain price range. The agency we worked with was very helpful, and we found the perfect place. This agency also had made some sort of listing mistake at one point, so we were kindly refunded fifty dollars! We appreciated this, as every dollar counted now more than ever because we were living in the more expensive United States. We were grateful for the churches and individuals who continued to support us in this new venture.

One afternoon, we looked at a house with a lovely picturesque Mimosa (silk) tree in the front yard. The owner was not there that day. We peeked in the front windows and then went around to the back to discover a large backyard surrounded by a privacy fence.

We just knew this was the house for us. Standing in that backyard, we prayed that we could be the ones to rent it. When we did meet with the owner, he told us that we were on the bottom of his list because of our financial information but on the top of his list in his character evaluation. He decided to take a chance on us, so we were in. Frankly, he had a deal too, because we believe that stewardship requires taking care of a rented house just as carefully as if we owned it. Also Oliver is an excellent "Mr. Fixit," and did repair jobs around the house labor-free, which the owner certainly appreciated. It was a happy arrangement.

Soon, the van arrived with our furniture from Minneapolis, and then we were pretty much settled in. Next, we sought ways to make ourselves known to the Vietnamese community, so we went to flea markets and Vietnamese stores. The fact that we spoke fluent Vietnamese certainly endeared us to the refugees.

We held special dinners at Thanksgiving and Christmas and put on programs for anyone interested. We also started teaching English to Asian immigrants. One time, we invited a family to go on a sightseeing trip. When we drove to meet them at the designated meeting place, many other cars were waiting for us. We ended up with a twelve-car caravan to the beautiful, sequoia trees of California at Calaveras Big Trees State Park. We enjoyed a picnic there together.

Knowing Vietnamese enabled us to listen to the frightful things many had experienced as they fled Communist Vietnam. The dangers these people faced on the high seas in small boats without adequate food and water was bad enough, but some of them had been intercepted, molested, and robbed by pirates. They told stories of terrible cruelty with tears running down their cheeks and ours, too. I was surprised by their expressions of heartfelt gratefulness for our help. Often, all we had done was listen. We then realized that they had tried to tell their stories to other Americans, but it was frustrating, as they did not yet know enough English and Americans did not understand Vietnamese. It was a unique opportunity for us and a tremendous blessing.

Our primary goal was to share the gospel with these people. We also wanted to help Christians grow in their Christian walk, so we held Bible studies every week in several homes. At first, all of our Bible studies were in Vietnamese. Later, the young people in the family wanted to study in English.

In the home of sister Van, Oliver taught her six daughters in English, while I taught the mother in another room in Vietnamese. Her husband, Loi, was still in Vietnam and had been in a re-education camp there for many years. We prayed for him for eight years, and what a happy day it was when he was able to join them in America. He, too, became a Christian. This particular Vietnamese Bible study group grew into a church, and we were glad when they called a Vietnamese pastor. That church is now a thriving evangelical Vietnamese church in Sacramento.

We witnessed many adjustments these reunited families had to make. The Vietnamese churches were wonderful anchors not only for their spiritual lives but also for their social lives as they struggled to retain some of their rich Vietnamese heritage while trying to assimilate here. The churches were a tremendous blessing and help to these families as they were getting adjusted, since many families had been torn apart and now, after many years, were together again.

Sometimes, in the early years of our ministry in California, our finances were rather low. Once we were given several packages of sausages, and it was quite challenging to discover how many tasty dishes I could make. Those sausages made wonderful stir fries, went well with noodles and cream of mushroom soup, were good cut up and added to scrambled eggs, delicious when wrapped in baking powder biscuit dough, and of course we had hot dogs. Since we never discussed our finances with people it was interesting that we were given things when someone had a surplus. The provision was always most welcome and especially so since it came just when there was a need. One day, a lady called us and said her mother periodically received supplies from a generous organization. She said her mother had too much stuff, and could she bring some of it over to us? She did, and it was a wonderful provision to us from the Lord through her.

In the early years, Oliver did quite a bit of interpreting in doctor's offices. He would be called at all hours to help with interpretation by phone in a three-way conversation between him, the patient, and the doctor. A Christian American lawyer needed Oliver's interpreting help in a court case. One day he came to visit us. After I handed him a mug of coffee, and we chatted, he suddenly turned to me and asked, "Do you have a car? I'll get you one," he said. And he did! I was flabbergasted! He was involved in auctioning cars. So that is how we got our 1973 Buick LeSabre.

One day, my brother Philip and his wife, Mary, came to visit us from Los Angeles. They took us to a Vietnamese restaurant, where we met a young Christian Vietnamese. Later, she was the cook in another restaurant and her non-Christian husband was the manager. Eventually, it was my privilege to drive to their restaurant once a week to have a one-on-one Bible study with her. She worked long hours every day of the week and was not allowed to attend church. She was one sad, lonely, young lady. We enjoyed our times together when her husband finally told her I was there and would allow her out of the kitchen for an hour.

Eventually, this couple had four children, the last three being only one year apart. Since the parents were so busy in the restaurant, the children were more or less left on their own in a back room. We became concerned when Daniel (Tam) was going to be sent to public school, as he knew virtually no English. He attended pre-kindergarten classes in a Christian school for a couple of months, where he suddenly realized there was a whole world out there, and that life was not just the restaurant and the back room. One day when we were visiting them, he laid his head on my lap and wailed in Vietnamese, "Oh, *Ba* (Mrs.) Joyce, pray that I can go to school!"

At that same time, we had learned from my sister Kathy of an organization that produced home school materials that were extremely inexpensive. After some discussion, Daniel's parents allowed us to home school him. We called our school Liberty Academy. Oliver picked him up every Tuesday morning along with his sister, Jadelynn (Phuong). Daniel was in kindergarten and Jadelynn in preschool.

We spent a lot of time teaching them basic English. For some time, they pronounced *tomorrow* as "tomollow." Also while learning English, they started to use the word *lousy*. Finally, we realized that they had combined the words *loud* and *noisy*. We always took them back to the restaurant on Friday night. This worked well. Eventually, the youngest daughter, Angeline (Vi), was old enough to start kindergarten. Later, she became my special helper, polishing my silver souvenir spoons that I inherited from my mother.

Homeschooling was a family effort for us. Jeanette had completed a year at Wheaton College, two years of Bible at Bethany, and was back home and working, as well as continuing her college studies. She was very generous in giving time to help the kids in their early English-learning efforts. She read books to them with many patient explanations.

Jonathan had a full-time job and was studying to earn his BA degree. He involved the kids in simple sport activities in our backyard and also taught them to play several board games. I recruited him to give them reading tests when they had finished a reading primer, which meant the children got to go to his room. How they loved that!

When they helped me make cookies, I would have one of the kids bring Jonathan some on a plate, and he loved that! Sometimes, though, he would creep out of his room with the express purpose of "stealing" some newly baked cookies. This put everyone on the alert, especially Jadelynn my baking helper, and when he appeared she would squeal "No, Jonathan, no!" Of course that did not stop him!

Oliver gave the children a Bible lesson each night and prayed with them. I was responsible for all their school subjects and we started each day with part of a Bible story, Scripture memorization, and songs. I also gave them beginning piano lessons after their mom bought us a used piano. It was rewarding watching them grow in academic and biblical knowledge and become good English speakers. I must add that Daniel brought his favorite rice bowl and his Vietnamese soup spoon when he first arrived. We smiled when one day he said in despair, "I can't remember how to say pineapple in Vietnamese!" We knew they were moving into our language and culture, and we were convinced that this was best for them as this is where they lived. They would always be able to understand and converse in Vietnamese. Our goal for them was that they become fluent in English.

There is something very enjoyable to me about being the one to teach a child to read, which had been my privilege with my own children and now with these three American-born Vietnamese children. We had many wonderful times together for six years. After that, they were enrolled in a fine Christian school to finish their elementary education and move on to the upper grades. During those years, we often had them stay with us for part of the summer months.

Oliver was eager to get certain tracts and other literature translated into Vietnamese. He worked on these with a Vietnamese Christian friend and produced and distributed them to individuals. He was especially happy when he got permission to translate into Vietnamese and produce one of his favorite tracts by Billy Graham, *Steps to Peace With God.* We handed these to any Vietnamese we met, and we still do so. Also, as other Vietnamese churches of various denominations came into existence in Sacramento, we supplied them with these tracts. We were involved in working with Vietnamese people and speaking Vietnamese, but as for our Hrey people, we had no contact with them whatsoever.

Imagine our utter joy when in 1993 we received a letter from Ngiah! It had been eighteen years since we last saw him when we said our farewells and left Vietnam and he headed back to his village at Bato. We were encouraged, too, when we opened the letter and found we had no trouble reading and understanding Hrey. Things had eased a bit in Vietnam so there was more open communication. We found a way to occasionally send him some much-needed financial aid.

Ngiah and E-Ne now had eight children. Later, it was such a joy to learn that their fifth child, a son, Em, had become a student pastor. We learned from Em himself that when he was a child he used to go around to the villages with his dad, who was witnessing whenever possible. He himself was not that interested, but what got his attention, he said, was his dad's faithfulness and that Ngiah occasionally received gifts from us. Em decided that this must mean his dad was doing something important and valuable. He too became a follower of Christ and was led into Christian ministry. He is now serving in the same valley where we once lived.

As some of the Vietnamese refugees became more fluent in English, we found we could then be of service to those whose language we did not speak. One man, a lay pastor from the Mien tribe in Laos, had come to Christ some years before in the mountains of Laos through the dedicated efforts of missionaries.

Jiu Choi would have Bible studies with us in English. Oliver supplied him with useful Christian literature. He asked lots of questions about different Scripture verses. His wife was a lovely Christian lady. Sometimes we were invited to their home and enjoyed yet another adventure in enjoying slightly different Asian meals, which always included steamed white rice. One thing I remember about this couple is that whenever they left our house, he marched ahead of his wife to their car. I could just imagine him marching along on a jungle trail, with her trailing behind him.

Jeanette received her medical assistant certificate with honors. After working as a medical assistant, she decided to do further studies and after graduating, got an excellent position in California and is now a happy, skillful, knowledgeable, successful, and appreciated certified physician assistant. We found it interesting that the military medics who had returned from Vietnam were skilled and experienced but were not fully trained doctors. The physician assistant program was originally set up to accommodate them and it became a popular opportunity for others.

Jonathan was moving forward in his educational requirements to become an ordained pastor. It is especially wonderful for parents to see their children work hard to become what they were obviously meant to become.

Unexpectedly, we found ourselves coming around full circle in our missionary careers. There was no way we could have anticipated that we would find ourselves back in the ministry that had been dearest to our hearts those many years ago. Yet that is exactly what happened!

Full Circle

Life for us in California had been blissfully busy as we served in various types of ministry. At first, we were there to meet refugees from Vietnam when they arrived at the Sacramento airport. We helped them find and get settled into apartments and were able to supply them with furniture, clothing, and other necessities donated by generous Christians.

We heard their heart-rending stories of escape from Vietnam and listened as they told of their difficult experiences on the long, wearisome, frightening trip in small fishing boats to freedom, often suffering from lack of food and water. The refugees finally arrived in Thailand or Indonesia, where they were placed in crowded refugee camps and waited many months while their papers were being processed.

When they flew into Sacramento, California, it was always interesting to observe Vietnamese families waiting to welcome a husband or father after years of separation. Usually, the men had been in harsh re-education camps. There were tears, smiles, shyness, excitement, and then untold adjustments for all concerned as they began a brand-new chapter in their family life together.

We continued with our ministry to the Vietnamese which included interpretation, visitation, translation, teaching English, preaching, Bible studies, producing tracts, and homeschooling. We were always available for the Vietnamese, and they knew it.

Gradually though, changes were taking place. Many Vietnamese were learning English and getting jobs or starting businesses. Families were growing up. Their children were graduating from college and getting married. Couples were moving from crowded apartments into nice homes in the suburbs. Refugees were becoming immigrants, and immigrants were becoming American citizens. Actually, my Australian-born husband became an American citizen on July 12, 1991, along with some of our Vietnamese acquaintances.

We watched the Vietnamese as they endeavored to adjust to life in America. Change is never easy, and this change was immense for many of them, though it was easier for some than for others. Their school children were quickly learning to speak English and learning American culture, which resulted in many changes in families. The children spoke Vietnamese at home and English at school. For some of the adults, especially the older ones, learning English seemed almost impossible. Therefore, our ability to speak Vietnamese continued to keep us busy in our ministry.

The life we once knew and loved with our beloved Hrey people had become only a fond memory. We had no contact with them whatsoever, except for an occasional letter from Ngiah. Then one day in 2003, we were joyfully stunned. Twenty-eight years after leaving Vietnam, we were invited to work as consultants for the translation of the Hrey New Testament. We continued our involvement in our Vietnamese ministry but now added part-time Hrey Bible translation to our list. The timing was perfect. I am ever amazed at God's provision but also awed and blessed by His timing.

By now, there were several Vietnamese churches in Sacramento. Capable, trained Vietnamese pastors began serving the Vietnamese community. Also, many government services were available to help new arrivals, and in addition to that, the Vietnamese had their own useful network of services. Things had changed a lot.

In 2005, God began nudging us that things were going to change for us too. Our house rent went up from $850 to $1,000 a month. Jeanette for some time had hoped to buy a house for us and decided to do just that. Then, right at that time, Jonathan decided to go to seminary in Texas to obtain his master of divinity degree. Texas was a much more economical place to buy a house than California at that time.

Oliver and I flew to Texas with the task of finding the perfect house for Jeanette to buy and for us to rent from her. We looked at several houses and ended up choosing the first one we saw, a four-bedroom split-level on a large corner lot. Ironically, there was a wooden cross that had been left on the wall of the dining room by the former owners, and it still hangs there.

Much discussion on the phone transpired between Jeanette and Jonathan in California and us in Texas. They were happy with our choice when they viewed all the pictures via the Internet. Oliver and I flew back to Sacramento. Before leaving the plane after landing, we received a call from the realtor in Texas, telling us that we (actually Jeanette) "have the house!" Everything was settled. That was January 14, 2006, which happened to be our forty-fifth wedding anniversary.

We were soon involved in a whirlwind of packing, and there were farewell parties and dinners for us, for which we felt honored. We had lived in Sacramento for twenty-four years. Oliver, Jonathan, his cat, and I left California in two cars. On March 6, 2006, we arrived in Spring, Texas. Our state, our residence, and even our ministry had changed. But what didn't change was God's faithful provision through faithful people who enable us to continue in this ministry. Oliver is occasionally invited to preach in Vietnamese in a local church, and more often to preach in English at their young adults and youth service. Otherwise, we are involved full time in Hrey Bible translation.

So that is how we came full circle to our original desire to get the Scriptures, the Old and New Testaments, into the hands of the Hrey people. We are working toward that goal. The wonder of instant communication by computer never ceases to impress us. It has certainly enhanced and expedited our Bible translation ministry.

Frankly, we are still somewhat surprised to find ourselves living in Texas. Jonathan met the lovely Sommer Bridges from South Carolina, a talented, accomplished, successful, Christian high school teacher here. She is definitely the one and only love of Jonathan's life and the answer to his and our prayers. They were married on October

4, 2008. We noted that our Jonathan met his Sommer in the winter in Spring, Texas, and they were married in the autumn.

One week after the wedding, Oliver and I were privileged to return to Vietnam for a visit, and were able to connect with many former Vietnamese friends, even a couple of Hrey men. It was a very special to have some time with them for Bible translation consultation.

Foreigners are absolutely forbidden to visit any tribal village in Vietnam. So imagine our joy when E-Ne and Em came to Danang to meet us. E-Ne almost did not come because she was terrified of riding on a bus! But she was finally encouraged by her son, who told her, "Don't worry, Mom. I will take care of you." She said she came, even though she was so scared, because she longed to see us. We gave her a small album of black-and-white photos of all of us back in our days together when our children were little. There was joy, and there were tears.

But we missed Ngiah, as he died in January 2005. We sat enthralled as Em told us that he was informed he would not be allowed to carry a cross in his father's funeral procession. This was too much. He told the officials that his father had served Christ for many years in that valley, and that he was going to have a cross in the procession. Actually, we received a picture, with someone leading the procession carrying a tall bamboo cross and several others carrying smaller ones. How delightful! And how brave of Em!

One October day in 2009, Oliver decided he wanted to buy me a dress for my birthday. *A dress? This is different,* I thought. I spruced up, put on high heels, and we went to our local discount store. Soon after we got there, he found a two-piece dress—red of course! I went into the dressing room to try it on and came out to show him. It was a perfect fit, and we both liked it.

Later, as we were browsing through the store, a young Christian woman with a small child in her shopping cart approached us. She had seen me come out of the dressing room and wanted to tell me she too thought the dress looked lovely. We learned that her husband was in the military overseas and she learned that we were missionaries. We then continued on our way to the checkout counter to pay for the dress.

Suddenly, that same young lady appeared again and insisted on paying for my dress. She absolutely insisted, and so she did! She told us she never shopped in that store, as she lived some distance away. She wished us well and walked out of our lives as we stood on the sidewalk outside the store, feeling quite stunned. We just looked at each other and said, "Thank You, Lord!" That was one of the most unusual experiences we have ever had in receiving an unexpected gift, and that from a complete stranger. It was a blessing. I wore the dress to a friend's wedding that same month.

Recognizing all the ordinary and sometimes unexpected ways in which the Lord blesses each of us every day is so refreshing. Recognizing Christ's presence and experiencing His beyond-understanding peace and presence, especially in times of difficulty, hardship, fear, grief, and trouble, is also incredible.

I love the story of the disciples out in their boat in a terrible storm on the Sea of Galilee. It must have been an incredibly severe storm for these experienced fishermen to be terrified. The situation was extremely serious, and Jesus was not with them. Suddenly, though, He appeared. He was walking on the storm-tossed water, which caused the disciples to become even more terrified. Was this a ghost? Then they heard Him say, "It is I; do not be afraid" (Matthew 14:27b). Even more incredible was that in the midst of their plight, He prefaced those words by saying, "Be of good cheer"!

We all experience storms in our lives when there is nothing we can do. We are helpless. We find ourselves bouncing up and down on powerful, wind-tossed waves in the darkness. Everyone sooner or later will be tossed about, even Christians. But every test is an opportunity to trust.

It is in those storms, in that time of need, grief, danger, pain, depression and uncertainty, that we need to listen carefully for those other incredibly comforting words of Jesus, "Lo, I am with you". Firmly believing that reality bolsters our faith and enables us to seize the situation as an opportunity to trust Him and experience His presence and help, not necessarily by feelings but rather by trusting in His presence in the person of God the Holy Spirit.

There is a special joy and blessing for all Christians who take careful note and are aware of God's personal, majestic, loving, quiet, intimate involvement and care in our lives, not just for finances but for a myriad of other things, if only we pause and pay attention. Sometimes the difficulty is prolonged, and yet even then, we can experience His presence and His real and ever-available peace, strength, and patience.

We are admonished to remember, "You shall remember all the ways which the Lord your God led you" (Deuteronomy 8:2a). Sometimes in the midst of our changing ministries, it was confusing to figure out what He was doing or going to do. That is why Scripture tells us to remember. By the simple act of looking back, we could more clearly trace His unfailing presence, gracious love, and gentle care. Remembering, emboldened us each time to dare to trust Him again as we stepped out into the unknown future.

All that we have experienced is our precious story of Christ's unfailing presence with us, as He promised. Each person who puts his or her faith in Christ as Savior and Lord will have a different story. However, each one will surely come to experience divine companionship, because Christ did not make an empty promise when He said, "Lo, I am with you" (Matthew 18:20b)

Ungaran, Indonesia

Indonesian language study

Madura, Indonesia

My DVBS

Oliver's TEE

Church in our living room

Our family—Indonesia

Bible study at sister Van's

A Church is born

Homeschooling—A Family Effort

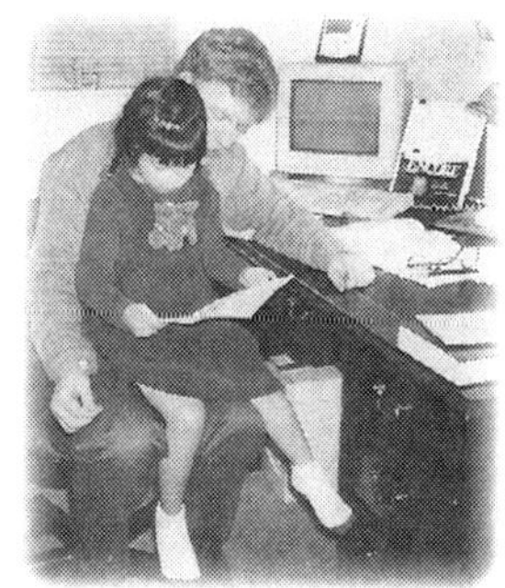

Jonathan/Sommer wedding

Full Circle: Bible translation continues

CPSIA information can be obtained at www.ICGtesting.com
Printed in the USA
BVOW07s1811171013

334031BV00001B/12/P